GUT FEMINISM

NEXT WAVE:

NEW DIRECTIONS IN

WOMEN'S STUDIES

A series edited by Inderpal Grewal,

Caren Kaplan, and Robyn Wiegman

GUT FEMINISM

ELIZABETH A. WILSON

DUKE UNIVERSITY PRESS

Durham and London 2015

Printed in the United States of
America on acid-free paper ∞
Designed by Amy Ruth Buchanan
Typeset in Quadraat and Gill Sans
by Westchester Publishing Services

Library of Congress Cataloging-in-Publication Data
Wilson, Elizabeth A. (Elizabeth Ann), [date]
Gut feminism / Elizabeth A. Wilson.
pages cm—(Next wave : new directions in women's studies)
Includes bibliographical references and index.
ISBN 978-0-8223-5951-7 (hardcover : alk. paper)
ISBN 978-0-8223-5970-8 (pbk. : alk. paper)
ISBN 978-0-8223-7520-3 (e-book)
1. Feminist theory. 2. Mind and body.
3. Feminism and science. 4. Depression,
Mental. I. Title. II. Series: Next wave.
HQ1190.W548 2015
305.4201—dc23
2015008324

Cover art: elin o'Hara slavick, *Global Economy* (slaughtered cow near Salvador-
Bahia, Brazil), 1996 (details). Chromogenic print. Courtesy of the artist.

Duke University Press gratefully acknowledges Emory
College of Arts and Sciences and the Laney Graduate
School, which provided funds toward the
publication of this book.

CONTENTS

ACKNOWLEDGMENTS

An early version of the second chapter of this book was published in 2004 in the journal *differences* under the title "Gut Feminism." In the short acknowledgments at the end of that essay I stated that this was the final expression of an argument made at greater length in my 2004 book *Psychosomatic: Feminism and the Neurological Body*. At the time, if I recall correctly, I imagined that the questions of the gut that had emerged late in the writing of *Psychosomatic* could be slightly extended, but that the *differences* article would bring those issues to a close. That is not what happened. In 2004 I was at the beginning of a five-year fellowship, funded by the Australian Research Council, on neurology and feminism. That project wasn't primarily oriented to questions of the gut, but as things have turned out, the gut and antidepressants have consumed all my attention. This book is the outcome of that research. The datum that 95 percent of the human body's serotonin can be found in the gut (something I first stumbled across while writing *Psychosomatic*) did not lose its grip on me. Exploiting these kinds of data, this book contains the best arguments I can currently muster for using the peripheral body to think psychologically, and for using depressive states to understand the necessary aggressions of feminist theory and politics. While feminist questions about biology and hostility will continue to be asked, I believe that I have now finally brought this particular, much extended project to a close.

There are many institutions and colleagues who have sustained me as I have written. I have been very fortunate to be invited to speak to a number of informed and animated audiences. Early versions of this research were presented at the following venues (in response to invitations from these colleagues): the Diane Weiss Memorial Lecture, Wesleyan University (Victoria Pitts-Taylor); the Linda Singer Memorial Lecture, Miami University (Gaile Pohlhaus); UCLA (Rachel Lee and Hannah Landecker); Kings College and the London School of Economics (Nikolas Rose); the University of California, San Diego (Lisa Cartwright and Steven Epstein); Concordia University (Marcie Frank); Women's Studies, Rutgers University (Belinda Edmondson); the Program in Women's Studies, Duke University (Ranji Khanna and Robyn Wiegman); the University of Illinois at Urbana–Champaign (Bruce Rosenstock); the Committee on Degrees in Studies of Women, Gender and Sexuality, Harvard University (Anne Fausto-Sterling); the Center for the Humanities, Wesleyan University (Robert Reynolds); the University of New South Wales (Vicki Kirby); St. Thomas Aquinas College (Charles Shepherdson); the MIT Program in Women's Studies (Evelyn Fox Keller); the Rock Ethics Institute and the Science, Medicine and Technology in Culture Program, Penn State University (Susan Squier); the Pembroke Center for Research and Teaching on Women, Brown University (Elizabeth Weed); the Centre for Women's Studies and Gender Research, Monash University (Maryanne Dever; JaneMaree Maher; Steven Angelides); the Department of Gender Studies, University of Sydney (Elspeth Probyn); the Australian Women's Studies Association; the Society for Literature and Science and the Arts.

This book was begun with the support of an Australian Research Council Fellowship (2004–2008) at the University of Sydney (Research Institute for Humanities and Social Sciences) and the University of New South Wales (School of English, Media and Performing Arts). The Australian Research Council has been an enormously important source of funding for me, and I remain very grateful for their support of interdisciplinary work that no doubt made them anxious. The project was also supported by a fellowship year at the Radcliffe Institute for Advanced Study, Harvard University (2011–2012), which provided me with remarkable resources and great intellectual company.

Some of the following chapters have been published in the early stages of this research; these essays have all been revised for this book. Chapter 1 appeared, in different form, as "Underbelly" in *differences: A Journal of Feminist Cultural Studies* 21, no. 1 (2010): 194–208. Chapter 2 was published, in different form, under the title "Gut feminism" in *differences: A Journal of Feminist Cultural Studies* 15, no. 3 (2004): 66–94. Fragments from "The work of antidepressants: Preliminary notes on how to build an alliance between feminism and psychopharmacology" (*BioSocieties: An Interdisciplinary Journal for the Social Studies of Life Sciences* 1 [2006], 125–131) and "Organic empathy: Feminism, psychopharmaceuticals and the embodiment of depression" (in Stacy Alaimo and Susan Hekman's *Material Feminisms* [Bloomington: Indiana University Press, 2009], 373–399) can be found scattered through chapter 4 and beyond. An earlier version of chapter 5 was published, in different form, with the title "Ingesting placebo" in *Australian Feminist Studies* 23 (2008): 31–42. Chapter 6 was published, in different form, as "Neurological entanglements: The case of pediatric depression, SSRIs and suicidal ideation" in *Subjectivity* 4, no. 3 (2011): 277–297.

I have the very best of colleagues in the Department of Women's, Gender, and Sexuality Studies at Emory University. My deepest thanks to two exemplary chairs (Lynne Huffer and Pamela Scully) who provided the conditions for this research to proceed, to a crack team of staff (Berky Abreu, April Biagioni, Linda Calloway, and Chelsea Long), and to my departmental colleagues Rizvana Bradley, Irene Browne, Michael Moon, Beth Reingold, Deboleena Roy, Holloway Sparks, and Rosemarie Garland Thomson. Special thanks to Carla Freeman, who provided excellent company and chocolate-based encouragement through a long summer when neither of us thought our books would ever be finished. Ingrid Meintjes helped with the dreary editing tasks right at the very end and was a lifesaver.

Over the many years of this book's formation my thinking has continued to grow in the company of great friends, supporters, and coconspirators: Steven Angelides, Karen Barad, Tyler Curtain, Guy Davidson, Penelope Deutscher, Richard Doyle, Anne Fausto-Sterling, Mike Fortun, Kim Fortun, Adam Frank, Jonathan Goldberg, Lynne Huffer, Annamarie Jagose, Lynne Joyrich, Helen Keane, Vicki Kirby, Neil Levi, Kate Livett, Elizabeth McMahon, Michael Moon, Brigitta Olubas, Isobel Pegrum, Marguerite Pigeon, Robert Reynolds, Jennifer Rutherford, Vanessa Smith, Colin Talley, Nicole Vitellone, and Elizabeth Weed. I would particularly like to note the importance of the Mrs. Klein reading group that met over many years in Sydney. My thanks to Sue Best, Gillian Straker, and kylie valentine not just for their incisive thinking and great humor but also for understanding that cake is necessary for sustained discussions of the Kleinian underworld. Robyn Wiegman provided a crucial reading of the manuscript in the latter stages that helped clarify what this book is about. I greatly appreciate her enthusiasm for intellectual and political adventure. Carla Freeman, Michael Moon, and Vanessa Smith read final snippets of the manuscript; they are the very best of friends and interlocutors. Courtney Berger has been a supportive and sanguine presence throughout the writing and publication process, and Erin Hanas, Liz Smith, and Christi Stanforth have very effectively guided the book into production. I remain very grateful for the support Duke University Press has shown to me over the years. Scott Conkright was an important, energizing influence in the final years of this project; his understanding that feeling and intellect are happy bedfellows repaired all kinds of problems. My New Zealand family have always been supportive of my intellectual endeavors. In particular, I have been especially grateful for a lively postal correspondence with Sarah Oram over many years. Ashley Shelden has a capacity for love that, daily, astonishes me. Let me say it as plainly as I can: she has brought me back to life. In relation to this book her intellectual acuity has been vital. Our conversations have strengthened and intensified the arguments presented here. My only wish is that the book could more ably reflect her abiding intellectual and emotional influence on me.

DEPRESSION, BIOLOGY,

AGGRESSION

The connections between gut and depression have been known, in the West, since ancient Greece. It was the Hippocratic writers who gave the name *melancholia* to states of dejection, hopelessness, and torpor. They understood such states to be caused by an accumulation of black bile (in Greek, *melaina chole*), a substance secreted by the liver. For these writers, and for practitioners of medicine for another two thousand years, melancholia was both the name of one of the enteric humors and the name for a disruption to emotional equilibrium (Jackson 1986). One of the Hippocratic aphorisms makes the affinity between these two modes of melancholia explicit: "The bowel should be treated in melancholics" (Hippocrates 1978, 217). The condensation of viscera and mood, exemplified in the term *melancholia*, is the subject of *Gut Feminism*. This book will explore the alliances of internal organs and minded states, not in relation to ancient texts but in the contemporary milieu where melancholias are organized as entanglements of affects, ideations, nerves, agitation, sociality, pills, and synaptic biochemistry.

I am not proposing a theory of depression. Rather, I want to extract from these analyses of depressed viscera and mood some gain for feminist theory. I have two ambitions. First, I seek some feminist theoretical gain in relation to how biological data can be used to think about minded and bodily states. What conceptual innovations would be possible if feminist theory wasn't so instinctively antibiological? Second, I seek some feminist theoretical gain in relation to thinking about the hostility (bile) intrinsic to our politics. What if feminist politics are necessarily more destructive than we are able to bear? This introduction offers some context for how feminist theory might approach these tough questions of biology and aggression.

In the first instance, this book makes an argument that biological data can be enormously helpful for feminist theory. By "helpful" I mean "arresting, transforming, taxing." When the project began (with a paper titled "Gut Feminism" at the 2003 Society for Literature and Science convention) my primary concern was to show that feminist theory could find conceptual insight in the biological and pharmacological research on depression. It had been clear to me for some time that there were significant gains to be made by reading biological, evolutionary, and cognitive research more closely (E. A. Wilson 1998, 2004). I wanted to show that data about the pharmaceutical treatment of depression need not always be the object of feminist suspicion; they could sometimes be the source of conceptual and methodological ingenuity. By 2003 many feminist science studies projects were expanding the ways in which biological data could be apprehended. In the wake of early influential work in feminist philosophy of science and biomedicine (e.g., Ruth Bleier, Donna Haraway, Sandra Harding, Emily Martin), Anne Fausto-Sterling's *Sexing the Body* (2000) and Evelyn Fox Keller's *Century of the Gene* (2000) brought to a wide audience the idea that biology was a site of important political and conceptual argumentation for feminism, and—in particular—that detailed understanding of biological processes was crucial to such feminist analyses. What followed were a number of important and engaging monographs on feminism, sex, gender, sexuality, capital, biotechnology, and biology: Susan Squier's *Liminal Lives* (2004), Catherine Waldby and Robert Mitchell's *Tissue Economies* (2006), Sarah Franklin's *Dolly Mixtures* (2007), Melinda Cooper's *Life as Surplus* (2008), Marsha Rosengarten's HIV *Interventions* (2009), Hannah Landecker's *Culturing Life* (2010), Rebecca Jordan-Young's *Brain Storm* (2010), Michelle Murphy's *Seizing the Means of Reproduction* (2012), Sarah Richardson's *Sex Itself* (2013)—to name those most prominent on my bookshelves over this decade.

My interests in *Gut Feminism* are less to do with that body of literature, which continues to flourish (and to provide sustenance for my own thinking), and more to do with the broader field of feminist theory, where biology remains something of a thorny conceptual and political issue and where antibiologism is still valued as currency. This book is less interested in what feminist theory might be able to say about biology than in what biology might be able to do for—do to—feminist theory. How do biological data arrest, transform, or tax the theoretical foundations of feminist theory?

Gut Feminism begins with the conjecture that despite the burgeoning work in feminist science studies there is still something about biology that remains troublesome for feminist theory. Take, for example, the feminist theoretical work on the body (which was very influential on my training and subsequent work). In the last thirty years, feminists have produced pioneering theories of the body—they have demonstrated how bodies vary across different cultural contexts and historical periods, how structures of gender and sexuality and race constitute bodies in very particular ways, how bodies are being fashioned by biomedical and technological invention. Yet despite its avowed interest in the body, this feminist work is often reluctant to engage directly with biological data. Most feminist research on the body has relied on the methods of social constructionism, which explore how cultural, social, symbolic, or linguistic constraints govern and sculpt the kinds of bodies we have. These theorists tend not to be very curious about the details of empirical claims in genetics, neurophysiology, evolutionary biology, pharmacology, or biochemistry.

This has been true even when biology is the topic at hand. Lynda Birke (2000), for example, provides a thorough overview of the early feminist work on the body. Like me, she is concerned that "the biological body has been peripheral to much feminist theory. . . . The emphasis in our theory was on the social construction of gender; the body hardly featured at all" (1–2). Like me, Birke expresses a desire to look inside the body, at the "blood and guts" (48). Nonetheless, and despite her training in neurophysiology and despite her desire to "bring the biological back to feminism" (175), Birke almost entirely avoids discussion of empirical data and focuses her analysis on the gendered narratives, metaphors, and representations that are "etched deep" (41) into biological knowledges. This aversion to biological data is widespread in feminist theories of all stripes. It bespeaks an ongoing discomfort with how to manage biological claims—as if biological data will overwhelm the ability of feminist theory to make cogent conceptual and political interventions.

One thing feminist theory still needs, even after decades of feminist work on the life sciences, is a conceptual toolkit for reading biology. In Psychosomatic: Feminism and the Neurological Body (2004) I thought at length about neurological data (the so-called gay brain, the neurophysiology of blushing, the peripheral neurology of neurosis) and their relation to feminist accounts of the body. However, I presumed too readily that lucid explication of biological detail would be enough to detach feminist theory from its conviction that social and discursive analysis are the primary or most powerful tools for engaging biological claims. My introduction to that book ends on a buoyant note: "It is the presumption of this book that sustained interest in biological detail will have a reorganizing effect on feminist theories of the body—that exploring the entanglements of biochemistry, affectivity, and the physiology of the internal organs will provide us with new avenues into the body. Attention to neurological detail . . . will enable feminist research to move past its dependency on social constructionism and generate more vibrant, biologically attuned account of the body" (Wilson 2004, 14). What I touched on in that book but did not pursue with any vigor was how important antibiologism has been to the successes of feminist theory. There is a powerful paradox in play: antibiologism both places significant conceptual limitations on feminist theory *and* has been one of the means by which feminist theory has prospered. Even as it restricts what feminist arguments can be made, antibiologism still wields the rhetorical power to make a feminist argument seem *right*. Because feminist theory has credentialed itself through these biological refusals, antibiologism is not something that can be easily relinquished. The opening two chapters of *Gut Feminism* tackle this problem directly. They describe this tendency to braid feminist theoretical innovation with antibiologism, and they discuss what legacies that leaves politically and conceptually. Because antibiologism has done such important authorizing work for feminist theory, any intervention that takes a nonparanoid approach to biological and pharmaceutical claims is likely to breach long-standing, dearly held feminist convictions. I anticipate that for many readers *Gut Feminism* is occasionally going to feel politically erroneous, dangerous, or compromised. This book takes that path, assuming that risk, in order to examine the tangle of antibiologism and critical sophistication that underwrites so much feminist argumentation.

As this project unfolded, another problem in relation to feminist theory and biology emerged. With the rise of the so-called neuroscientific turn in the critical humanities and social sciences in the last decade (Fitzgerald and Callard 2014; Littlefield and Johnson 2012), feminists and other critics began to take biological claims more seriously. However, they have often done so in a way that was overly credulous about the status of neuroscientific data. This is the coin of antibiologism flipped verso. Where traditionally many feminists have preemptively dismissed biological claims, this new breed of neurologically informed critics want to swallow biological claims whole: "We are living at the hour of neuronal liberation" (Malabou 2008, 8). In analyses like this, engagement with biology has more often meant betrothal than battle. *Gut Feminism* will intervene into this broad problematic (not enough engagement with biology; too much belief in biology) by reading for what is peripheral in biological and pharmacological theories of depression, and for what the psychoanalyst Sándor Ferenczi called the biological unconscious. By focusing on the neurological periphery (the enteric nervous system that encases the gut) I aim to show that biology is much more dynamic than feminists have presumed and much less determinate than many neuro-critics currently suppose. Specifically, this book contests the idea that neurological arguments are always about the central nervous system (the brain, the spinal cord): the neurological is not synonymous with the cerebral. This is one place, it seems to me, where the new neuro-critics have been too compliant with the convention that the neurology that counts is all above the neck. I want to show how some biological and pharmacological data about depression help us think about minded states as enacted not just by the brain but also by the distributed network of nerves that innervates the periphery (especially the gut). My argument is not that the gut *contributes* to minded states, but that the gut *is* an organ of mind: it ruminates, deliberates, comprehends.

These concerns about how to read biology were the first and explicit goal of *Gut Feminism*. These were the key problems that I researched and intended to analyze. The second major consideration of this book emerged from the presentation, revision, and rereading of the manuscript, and it is not something that I had anticipated in the early parts of the project: I found myself making a strong case for the necessary place of aggression (bile) in feminist theory. There are some obvious intellectual antecedents for such a claim (feminist anger; deconstruction; Kleinian psychoanalysis), but the most prominent of these for me has been the so-called antisocial thesis in queer theory. In the latter stages of this project I have been teaching the now canonical queer work of Leo Bersani and Lee Edelman, and I have had to work especially hard against the tendency in students to read self-shattering or negativity as an apolitical force that works to simply undo the coherence of the social or the subject. What Bersani and Edelman propose is not the punk sentiment that wants "to fail, to make a mess, to fuck shit up" (Halberstam 2006, 824), a sentiment that speaks only to consciously accessible parts of the social fabric and that sees negativity only in the realm of rebellion and antinormativity. One important pedagogical goal of these classes has been to make clear that negativity is intrinsic (rather than antagonistic) to sociality and subjectivity (Berlant and Edelman 2013), and this makes a world of difference politically. This queer work isn't antisocial at all; rather, it wants to build theories that can stomach the fundamental involvement of negativity in sociality and subjectivity.

The idea that negativity is indeed *negative* has been a hard lesson to learn. Today (Friday, June 13, 2014), as I sit down to rewrite this introduction, there is a one-day feminist and queer event called "Radical Negativity" at Goldsmiths College, University of London (http://radicalnegativity .com). The byline for this event encapsulates a widespread conceptual problem with how to approach negativity and aggression. The conference website describes the event as "an interdisciplinary conference interrogating productive possibilities for negative states of being," and the description of the conference describes a shared hope to "valorise negative states" in order to "provide the potential to open up new possibilities for politics and connection." Against this idea that the negative can be made valuable (productive, valorized, connected), *Gut Feminism* makes a case that we need to pay more attention to the destructive and damaging aspects of politics that cannot be repurposed to good ends. Chapter 3 takes up this argument in depth. There I claim not only that depression is a more outwardly aggressive event than we usually think (it is not just the inward turn of aggression against oneself), but also that this outward turn of hostility is the mark of every political action. In important, unavoidable ways, feminist politics attack and damage the things they love. This encounter with a negativity *that stays negative* continues to be an important thread through chapters 4, 5, and 6, where the particulars of antidepressant treatment are examined. Feminist politics are most effective, I argue, not when they transform the destructive into the productive, but when they are able to tolerate their own capacity for harm.

What is the nature of contemporary melancholia? The OED notes that the use of the word *melancholy* to denote ill temper, sullenness, and anger generated by black bile is obsolete. Contemporary melancholic states are less humoral, more molecular. Since 1987, when fluoxetine hydrochloride (Prozac) was approved by the US Food and Drug Administration (FDA) for use as an antidepressant, melancholias in the United States and its pharmaceutical outposts have been significantly transformed. They have become more prevalent, more quotidian, more biochemical, more cerebral, and (paradoxically) both more treatable and more intractable. There are some stable demographics to these post-Prozac depressions: they tend to be diagnosed more frequently in women than in men, and worldwide rates of depression are higher in the poor than in the wealthy. Yet melancholia has always been characterized more by mutability than by a stable set of symptoms. For example, early psychoanalytic theories of melancholia were oriented toward bipolar conditions (the circular insanities), and these theories were more attentive to the significatory patterns in melancholic states than would be the case in the second, medicated half of the twentieth century. These first psychoanalytic patients were despondent in particular kinds of ways: cannibalistic, ambivalent, lost (Abraham 1911; Freud 1917a). After imipramine (the first tricyclic antidepressant, synthesized in the 1950s) and with the subsequent revisions of the *Diagnostic and Statistical Manual of Mental Disorders* (DSM) away from Freudian and toward biological etiologies, depressions began to look and feel different. The identification and diagnosis of depression in medical environments underwent significant revision (Healy 1997), and the form and experience of our depressions changed accordingly. Despite being more biochemical, these later depressions were less vegetative (somatic) in character; and in the wake of Aaron Beck's (1967) influential cognitive theory of depression, depressions became more ideational in tone—the result of distorted thinking rather than disordered feeling or imagination or libido.

These transformations are only the most recent change in a long history of metamorphoses: in prior centuries melancholia was variously a sin, a madness or monomania, a form of love, an unabating fear, the stagnation of blood, a delirium of the brain, a dotage without fever, a temperamental trait, or degeneration (Radden 2000). Robert Burton's

Anatomy of Melancholy (1621/1989) famously finds not one, but seemingly endless causes, symptoms, and forms of melancholic distress. In fact, much of what we have come to take for granted about depression has been historically variable: the association of depression with women, for example, is a fairly recent event—it emerges after a very long history (from Aristotle to Hamlet to Churchill) in which melancholia was a gauge of male genius (Schiesari 1992). Moreover, even in the current milieu there are significant variations in how depression is experienced and treated: Arthur Kleinman found culturally distinctive links between somatic symptomology and neurasthenia and depression in China (Kleinman 1986), and one of the most highly industrialized countries in the world, Japan, was unusually slow in taking up new antidepressant pharmaceuticals (Berger and Fukunishi 1996; Kirmayer 2002). Intrasocietal differences add to the complexity: for example, individuals of East Asian descent living in North America, Europe, or Australia are less likely than individuals of European descent to develop major depressive disorder (as described by the DSM), despite the likelihood that they will experience more discrimination, a lower standard of living, and poorer medical care (Dutton 2009).

Melancholia, then, finds no one form across time and place. This mutability is crucial to the arguments in *Gut Feminism*. The conceptual, political, and methodological orientations of this project emerge from the presumption that depression is *contingent*. Even though melancholia has endured from ancient Greece until the present, it is not a condition underwritten by substrata that persist despite variation in historical, cultural, or discursive positioning. Such foundations (too easily designated "biological") simply do not exist in that form. But neither is melancholia just the effect of cultural trends, a fad (or weak ideological construct) that might be dispelled by incisive critique. This project tries to demonstrate the inadequacy of thinking in such bifurcated terms (Is depression biological or is it socially constructed? Is depression created by biochemical imbalances or historical injustices?). Those nature/culture debates presume a separation in the substrata of depression, as if biochemicals and cultural institutions are oil and water. *Gut Feminism* disputes the compartmentalization of nature/culture arguments in all their forms. It will be my presumption throughout that biology and culture are not separate, agonistic forces; that a political choice cannot be made between biological and cultural agency; that the interaction of biology and culture (nature/nurture) is an inadequate solution to the problem of etiology; that the relative weight of biological or cultural factors cannot be individually calculated; that biology is not a synonym for determinism and sociality is not a synonym for transformation. I use "contingency" here (borrowed from Barbara Herrnstein Smith [1988]) to mark that state of nature-culture entanglement that is almost impossible to articulate: coimplication, coevolution, mutuality, intra-action, dynamic systems, embeddedness (names given by the feminist philosophers of science whose work is foundational to this book's claims: Karen Barad [2007], Anne Fausto-Sterling, Cynthia Garcia Coll, and Megan Lamarre [2012a, 2012b], Evelyn Fox Keller [2010], Susan Oyama [2000]).

There are contingencies to melancholia that stretch across two thousand years, and there are contingencies that have found form only in the wake of certain pharmaceutical interventions since the 1950s. *Gut Feminism* is concerned primarily with the biological and pharmaceutical contingencies that shape depressions after Prozac. If the ways in which depressions are diagnosed, experienced, and treated are all intralinked (that is, if how we talk about depressions, and how we treat them, and how they crystallize biologically, and what they feel like, are all mutually coimplicated), and if the pattern of this mutuality has been subject to recent transformation, then my critical approach to understanding the pharmaceutical treatment of depression must be able to adapt to these new circumstances. In particular, this project presumes that the political commitments of antipsychiatry that flourished in the period between the first generation of antidepressants in the 1950s and the arrival of the selective serotonin reuptake inhibitors (SSRIs) in the 1990s are usually ill suited for the current melancholic landscape.

There are a number of reasons that antipsychiatric politics have less purchase now than one might expect. In the first instance, the deinstitutionalization of psychiatric patients that began in the United States, the United Kingdom, Europe, Australia, and New Zealand from the 1960s onward led to significant decreases in people incarcerated in psychiatric care (Fakhoury and Priebe 2002). For example, in 1955 in the United States (as antidepressants and antipsychotic drugs were first coming on the market), 559,000 people were institutionalized in state mental hospitals; by 1971 this number had dropped to 275,000, and by 1994 this number had dropped again to 72,000 (Bachrach 1976;

Lamb 1998). The politics of the asylum—so cogently pursued by R. D. Laing and others—have less relevance as psychiatric inmates become outpatients. Attempts to apply antipsychiatric politics to post-Prozac depressions (e.g., Breggin and Breggin 1994; Cvetkovich 2012; Davis 2013; Griggers 1997, 1998) have been much less compelling than earlier work in large part because the most intensive sites of depressive action in these deinstitutionalized countries are no longer hospital wards. Instead depressions have become extensively entangled with everyday life: support groups, talk shows, memoir, self-help books, op-ed pages, personal anecdote, blogs, social media, direct-to-consumer advertising. Analysis of these phenomena doesn't follow directly from critiques of institutionalized psychiatry.

Second, the prescription of antidepressants has moved out of the hands of psychiatric specialists in hospitals and into the hands of different kinds of health practitioners. In Australia in 2000, for example, 86 percent of prescriptions for antidepressants were made by general practitioners, who do not have extensive psychiatric or psychotherapeutic training (McManus et al. 2003). In the United States, nurse practitioners and (in some states) clinical psychologists are able to prescribe antidepressant medications (Shell 2001). For better or for worse, the treatment of depression is no longer under the direct control of psychiatric expertise. Few of the sources that I turn to in this project are psychiatric; instead, Gut Feminism focuses on psychological and psychoanalytic theories of depression, especially as they relate to the use of pharmaceuticals and as they engage biological data. While psychological and psychoanalytic research increasingly draws on biological (usually neurological) data, it is not primarily focused on biological explanations and is not wedded to the disease model of depression that exemplifies psychiatry. This means that biological data in psychological and psychoanalytic theories are often less sequestered from interpersonal, intrapsychic, social, economic, or historical events; as such, these theories are often well positioned to engage the logic of entanglement that I am pursuing here.

The third reason antipsychiatry has less leverage than it had in the past is that there has been massive growth in the population in deinstitutionalized countries that consumes antidepressant medications. Nikolas Rose (2003) calculated that prescriptions for SSRI pharmaceuticals in the United States increased by 1,300 percent from 1990 to 2000. It is my hunch that what underwrites the increase of antidepressant use is neither a malevolent intensification of psychiatric influence nor a miraculous leap forward in terms of pharmaceutical efficacy (the new-generation antidepressants are about as clinically effective as the older MAOIs and tricyclic drugs). In large part, what made these new drugs more palatable to patient and practitioner alike is a decrease in aversive side effects. Tricyclic antidepressants (like imipramine) can provoke a variety of distressing side effects (dry mouth, blurred vision, constipation, dizziness, sedation, weight gain), and the MAOI antidepressants force patients into a strict diet, as some common foods (e.g., chicken and cheese) can react fatally with the medication. When Prozac was released into the US market, it promised effective symptom relief with few side effects. For this reason SSRI pharmaceuticals were prescribed to people who had more minor forms of depression (e.g., what was then known as dysthymia, chronic and debilitating low mood)—conditions that previously would not have been considered serious enough to warrant psychiatric medications and their possible adverse effects.

It is not clear, then, that corporate and biomedical duress have been the most important mechanism for the expansion of antidepressant use, or that struggles over corporate and biomedical malfeasance (crucial as these battles are) generate the best avenue of approach to the treatment of contemporary depressive states. I am hypothesizing that if recent shifts in the treatment of depression have been made not on the back of an antidepressant's therapeutic effect, but on the basis of its so-called secondary effects—effects that pharmaceutical companies and medical practitioners and patients alike attempt to minimize—then questions of political and biological influence as they have been thought in antipsychiatry need to be comprehensively reconsidered. A drug's side effects run athwart not just therapeutic effects but also the institutions and the personal treatment regimes that attempt to manage drug efficacy. For this reason Gut Feminism will be intensely interested in the agency of what is allegedly ancillary to pharmaceutical action (e.g., placebo). Chapters 5 and 6 will expand these claims in relation to placebo and pediatric use of antidepressant medication.

It is my feeling, then, that the insurrection that powers antipsychiatry is mismatched to the post-Prozac landscape. This mismatch has been particularly debilitating for feminist critique. Despite a robust women's health movement dating back to the late 1960s (Morgen 2002) and despite intensive criticism of pharmaceuticals like Viagra (Tiefer 2010), there has been surprisingly little feminist commentary on Prozac and its sibling pharmaceuticals. Judith Kegan Gardiner (1995) begins a review of *Listening to Prozac, Talking Back to Prozac,* and *Prozac Nation* with an anecdote that encapsulates this impasse between the popularity of pharmaceutical treatments on the one hand and the preference for antipsychiatric (socially constructed) theories of depression on the other:

> I recently attended an interdisciplinary feminist meeting that assumed a consensus about social constructionism and criticized scholarly work that was perceived as "essentialist," because it implied a biological basis for gender attributes. During meals and breaks, however, I heard a different story. Several women were taking Prozac or similar drugs for depression. Some of their children, who had been difficult, "underachieving," or disruptive in school, were also being medicated. These informal discussions centered on symptoms, side effects, and relief. They implied but did not discuss a view of personality as biochemically influenced. . . . The potential contradiction between such private solutions and the publically avowed ideology of social constructionism was never voiced. (Gardiner 1995, 501–502)

It seems that our best accounts of psychological politics have developed independently of biological data, and also—if we accept the veracity of Gardiner's story—independently of everyday life. Somewhat surprisingly, this deficit is also visible in the cultural commentary on depression and loss (where the ubiquitous use of antidepressants might be the object of critical investigation): the leading edge of work on melancholia in the critical humanities is largely mute about the questions of pharmaceutical use and pharmacological action (e.g., Eng and Kazanjian 2003). The conundrum that underwrites *Gut Feminism* is this: how to engage pharmaceutical data about the treatment of depression without simply acquiescing to the idea that biomedical research provides the factual foundation on which interpretations are built (on the one hand) or routinely repeating the doxa of social constructionism (on the other). My method is to take contemporary biomedical data about depression and read them through the peripheral body—specifically, the gut: that is, to push the serotonin hypothesis about depression out past the central nervous system, and likely out past the limits of its coherence. I will take the biomedical data seriously but not literally, moving them outside the zones of interpretive comfort that they usually occupy. My goal is to draw in the periphery of the body as psychological substrate and deisolate brain from body, psyche from chemical, neuron from world. In concert with (and greatly indebted to) the work of Karen Barad (2007) and Vicki Kirby (1997, 2011), I am interested in the entanglements and patternment that generate depressive states. To this end *Gut Feminism* is not an argument for the gut *over* the brain, for drugs *instead of* talk, for biology *but not* culture; rather, it is an exploration of the remarkable intra-actions of melancholic and pharmaceutical events in the human body.

....

Let me give an example of how I will read biological data in the chapters that follow: abdominal migraine. The name itself is something of a catachresis—a condition ordinarily known to be in the head has found its way to the gut. Overseer and underbelly are already confused. The clinical characteristics of abdominal migraine are somewhat diffuse. It is commonest in children, and is characterized by acute and incapacitating midline abdominal pain, lasting hours or perhaps even days. The pain is dull and poorly localized in the epigastric or umbilical regions; attacks are recurrent. These days abdominal migraine is classified as one of the functional pediatric gastrointestinal disorders, meaning that (along with infant rumination, cyclic vomiting, and irritable bowel syndrome in children) the condition has no known organic etiology (Rasquin-Weber et al. 1999). When afflicted the child will usually exhibit a deathly pallor and will shun food; she or he may also have the symptoms of classical migraine (prodromal aura, headache, photophobia, vomiting). In between attacks the child is healthy. The symptoms tend to diminish as the child grows and many children spontaneously recover after a few years, although these individuals may be more prone to migraine as adults (Dignan, Abu-Arafeh, and Russell 2001).

There is, unsurprisingly, some disagreement in the clinical literature about whether abdominal migraine ought to be thought of as a psychogenic affliction or as a condition caused by biochemical abnormalities. The biologically oriented studies tend to think of abdominal migraine as a "migraine equivalent" (Dignan, Abu-Arafeh, and Russell

2001, 55): it is a classic head migraine, but with variant symptomology. Like classic migraine, these researchers argue, it is the dynamics of vasoconstriction/vasodilation and neurotransmitter irregularities—all located in the central nervous system—that underpin the symptoms of childhood abdominal migraine. This literature thinks of abdominal migraine as a cerebral attack, exiled to the periphery. What I will argue in *Gut Feminism* is not that the periphery is a site of abandonment (a maligned fringe, a desolate border), but rather that the periphery is a site of intense biological, pharmaceutical, and psychological agency on which the center is always vitally dependent. Which is to say, the periphery is interior to the center; the stomach is intrinsic to mind.

Sometimes the biological studies about abdominal migraine hint, inadvertently, at this kind of tangle between psyche and soma, head and gut. David Symon and George Russell (1986), for example, are clear that abdominal migraine should not be confused with a psychogenic condition; they see it as a "distinct clinical entity" that can be "easily distinguished" (226) from other patterns of childhood abdominal pain. They see abdominal migraine as more head-y, less enteric, than its name implies. They support this hypothesis with a treatment study of children with abdominal migraine. Twenty children were treated prophylactically with the drug pizotifen, an antimigraine medication, for two to six months: 70 percent experienced complete remission of their symptoms, compared with 15 percent in the control group who "received no treatment other than explanation and reassurance" (225). Symon and Russell conclude that these data provide evidence that abdominal migraine is less a psychogenic condition (that would respond to the palliatives of explanation and reassurance) and more like classic migraine, seemingly a disease whose etiology can be located above the neck.

These data are very persuasive. The etiology of abdominal migraine seems to be heavily weighted toward biology. But let's keep looking. What is the nature of the drug with which Symon and Russell treated these children? Pizotifen is a serotonergic antagonist. The logic that underlies its use in this study is that it inhibits the action of certain serotonin receptors, thus blocking the effects of serotonin in the nervous system that precede classic migraine attacks (vasoconstriction and vasodilation). Symon and Russell give no specific bodily location for pizotifen's effects; but their citation of several established studies about the drug's cognitive efficacy, along with their lack of interest in peripheral neurology, imply that they see the neurology of abdominal migraine as centrally (cerebrally) controlled. We might call this "listening to pizotifen" (Kramer 1993): if the condition responds to an antimigraine treatment then, properly speaking, it is a migraine of the head.

However, some of the studies that Symon and Russell cite to support their use of pizotifen as a treatment suggest a more distributed, less neurologically circumscribed character to migraine—abdominal and otherwise. For example, one of their citational sources (Hsu et al. 1977) found increased levels of catecholamines (specifically, noradrenalin) in the blood plasma of migraine sufferers. Noradrenalin acts as a neurotransmitter in the human nervous system, so it tends to circulate discursively as a sign of central nervous action, further consolidating the cerebral reputation of migraine. However, while noradrenalin is released by a region in the brainstem (the locus coeruleus) in response to stress and acts on a variety of brain structures (amygdala, hippocampus, neocortex), so too is it released by the peripheral (sympathetic) nervous system and the adrenal glands as part of the body's fight-or-flight response. This means that noradrenalin is a chemical generated by, and widely distributed through, the central and peripheral nervous systems of the human body. Elevated levels of noradrenalin in the plasma of migraineurs (extracted from a vein in the arm) indicate body-wide neurotransmission in migrainous attacks.

These data no more indicate the authority of the central nervous system over the peripheral nervous system than they do the reverse. Abdominal migraine is no more a variant of classical cerebral migraine than classical migraine is a deviation from abdominal pain. In both of these studies, the locus of the migraine remains an open question—unexplored in Symon and Russell, and clearly dislodged from its conventional home inside the cranium by Hsu et al. (who see a wider network of migrainous agency: plasma–sleep–stress–noradrenaline–personality–somatization–nutrition–diet–agitated mood). My suggestion is not that the question of migraine's character has yet to be empirically resolved (is it of the head or of the gut?), but rather that the character of migraine is truly open and distributed. Situated athwart our usual expectations that the biology of mind will be central and locatable, migraine is an engaging conceptual object for feminist theory.

An older study (that speaks more directly to the psychological and perhaps depressive aspects of abdominal migraine) helps elucidate what is at stake conceptually. Farquhar (1956) summarizes 112 cases of abdominal migraine in children and documents the emotional state of children with abdominal migraine (data that tend to drop out of more recent literatures): "worriers," "obsessional," "nervous" (1084). He also notes that "the relationship between migraine and liver dysfunction has long been recognized" (1084). This comment is made in relation to migraine and diet (fatty foods may aggravate abdominal pain in these children), and could be read simply as an organic aside. However, the question of the liver connects elsewhere in his paper to one of the most common manifestations of abdominal migraine in children: biliousness. In these literatures, bilious is sometimes simply a synonym for abdominal migraine ("bilious attacks" [1082]). However, the term (like the humor from which it emerges etymologically) also refers to emotional states—peevishness and ill-temper (the OED defines bilious equally as "of diseases and temperament"). Biliousness, in the context of abdominal migraine, is undecidably both of the liver and of the mind—a condensation of ailment and disposition. Like melancholia, its humoral cousin, a bilious attack brings our attention to the psychic character of the gut and to the enteric character of mood.

Abdominal migraine encapsulates the problems that feminist theory encounters when it reads biology. The interpretation of biological data in biomedical research often splits mind from body, locates mind only in the brain, and so thinks of the biological periphery as psychologically inert. It is also difficult to know what to do conceptually with bile and aggression. The next three chapters take on these issues in detail. They argue that depressive rumination is as visceral as it is ideational, and they track some circuits of hostility in feminist politics. With these analyses of antibiologism and aggression in hand, the final three chapters focus on the efficacies and failures of antidepressant medications. These final chapters outline how SSRI medications work biologically, and they explore one of the core contradictions in critical accounts of the pharmaceutical treatment of depression—that SSRIs are both ineffective (no better than placebo) and harmful (especially to children and adolescents). These twin problematics (biology and aggression) structure the arguments that follow. In relation to depression, I argue that feminist theory could engage the contemporary landscape more potently if it was able to read biology more closely and tolerate the capacity for harm. While the focus of the book is the pharmaceutical treatment of depression, I am hoping that a general orientation to the reading of biology and hostility emerges from these pages. I have provisionally called this method "gut feminism"—a feminist theory that is able to think innovatively and organically at the same time.

PART I

FEMINIST

THEORY

CHAPTER I

UNDERBELLY

Let me begin with a clinical vignette that will get us oriented toward the gut and depression and feminist theory. This fragment is drawn from the work of Darian Leader—a Lacanian analyst working in London who also writes popular books about psychology and is an occasional columnist for the *Guardian* newspaper. His 2008 book on depression opens with this sketch:

> After receiving a prescription for one of the most popular antidepressant drugs on the market and picking them up from her chemist, a young woman returned home and opened the small packet. She had imagined a yellowish bottle filled with tightly packed capsules, like vitamin pills. Instead she found flat metallic wrapping, with each pill separated from its neighbour by a disproportionate expanse of empty foil. "Each pill is in total solitude," she said, "like in metal shells looking out at each other. They are all in individual prisons. Why aren't they all together in one box, loose and free?" The way the pills were packaged troubled her. "They are aligned like obedient little soldiers—why doesn't at least one of them break rank?" Her next thought was to swallow all the pills together. When I asked her why, she said, "So they don't feel so lonely and claustrophobic." (Leader 2008, 1)

There are a number of different interpretive paths we could take in relation to this anecdote. We could inquire into the young woman's identification with the solitude of the pills, her phantasy that they ought to break free from their regimented existence, her nostalgia for medications that come in yellowish bottles rather than machine-pressed, mass-produced packaging, and her expectation that antidepressants and vitamins might somehow be akin.[1] All of these paths would take us into familiar clinical territory, where the young woman's associations could be analyzed within a well-established set of rules about neurotic ideation. Or, following Leader, we might be interested in the broader cultural significance of the story: "We could see this situation as a metaphor for the way that depression is so often treated in today's society. The interior life of the sufferer is left unexamined, and priority given to medicalizing solutions" (2). That is, we could deploy this story as a way to push back against mechanized, biologistic, market-driven treatments of depression, and demand (as Leader does) that more attention be given to the complexity of unconscious mental life.

Both these approaches are important, but I want to get at something else in this story. I am struck that the patient thought of *swallowing* the pills—all of them, together. The young woman doesn't look for one of those yellowish bottles into which she could put the pills. Rather, she feels that their loneliness and claustrophobia would be alleviated by being in her stomach—as if she understands that her gut and the psychic world are allied. Might there be more than one kind of "interior life" being documented here? Psychological and psychoanalytic and psychiatric theories of depression draw on cognitive distortion, unconscious motivation, or neurochemical imbalance; rarely do they talk about bodily action in such a direct way. I open with this anecdote not in order to suggest that the body underwrites depression—that it might be the substrate that causes cognitive or affective dysfunction. Rather, I want to think anew about the character of biology in such research; in this chapter I will argue that phantasy and peristalsis (swallowing) are coeval. That is, the gut is always minded: it ruminates.

In these first three chapters, I want to think about the rudimentary processes of ingestion, digestion, peristalsis, and vomiting as part of the psychic landscape; and I want to bring feminist readings of depression into closer contact with these alimentary events. Part of my difficulty in reading gut, feminism, and depression together is that the sophisticated feminist theoretical work that would normally sustain this kind of project often takes its distance from rudimentary bodily processes. Curiously, eating and hunger often figure in feminist work as the mark against which more cultivated critical and political stances can be built. In this chapter I will concentrate on one example (Gayle Rubin's influential work on gender and sexuality) that illustrates how feminism has placed biology at a distance from its own conceptual and political affairs. The implications of this for feminist theory are canvassed, and some suggestions, via Melanie Klein, are offered as a way to think about the psychic nature of the organic interior. In the two chapters that follow, I build on these preliminary ideas to think about a biological unconscious (chapter 2) and about the nature of psychic and political aggression (chapter 3).

Bringing Up Biology

Rubin's two canonical essays, "The Traffic in Women" (1975) and "Thinking Sex" (1984), were watershed moments for feminist theory. In her interview with Rubin in 1994, Judith Butler begins by noting that with these essays, Rubin "set the methodology for feminist theory, then the methodology for lesbian and gay studies" (Rubin in Butler 1994, 62). And indeed, it is clear that Rubin's work was influential for some of the most important theorists of gender and sexuality who emerged in the generation following, including Butler herself. If we were to think of Rubin's contributions to feminist theory in axiomatic form, we could say that in 1975 she argued for a disarticulation of biological sex from gender and in 1984 she argued for a disarticulation of the study of gender from the study of sexuality. In the 1994 interview, Rubin demurs from the presumption that she instigated these theoretical and political changes. Instead, she notes that both essays emerged out of an already existing set of concerns in her political and intellectual communities: the lack of an adequate analysis of gender in Marxism (in the 1970s) and the rise of antisex feminism (in the 1980s). Her essays were able to set a new methodological tone because they clearly articulated a conceptual shift that her feminist cohort already craved.

Rubin also mentions in the 1994 interview that there is no direct line connecting "The Traffic in Women" to "Thinking Sex." Her political concerns in 1984 had arisen in a manner somewhat orthogonal to those that had motivated her in 1975: "I was trying to get at something different" (67). The sometimes stormy relation between a politics of gender and a politics of sexuality that followed from "Thinking Sex" (Halley 2004; Wiegman 2004) also tends to accentuate the ways these two essays, and their two constituencies, can be set apart. In this chapter I travel along a different axis of analysis, one that binds "The Traffic in Women" and "Thinking Sex" more closely together. It is Rubin's orientation to biological explanation (or, rather, her turn away from biological explanation) that interests me here. I will argue that, despite their differences, these essays share a common commitment in relation to biological substrata and politics; both essays argue that biology doesn't have much to do with politics, or at least it has no constructive bearing on politics. I am turning to Rubin in order to explore one route by which biology became the underbelly of feminist theory: how it became both a dank, disreputable mode of explanation and a site of political vulnerability. By examining the dynamics of antibiologism in Rubin's influential essays, I am hoping to broaden the base of what can count as theory and what can count as feminist innovation.

My argument focuses around a set of claims about biology and the belly in "Thinking Sex." Rubin's assertions about how to handle biological theories of sexuality emerge as she situates her work (and the work of people like Michel Foucault and Jeffrey Weeks) as "an alternative to sexual essentialism" (Rubin 1984, 276). The specific form of essentialism that she targets is the idea that sex and sexuality are natural forms (i.e., fixed biological or psychological types) that exist prior to social life. Rubin rejects this formulation of sex and sexuality as her first matter of business: sexuality, she argues, "is constituted in society and history, not biologically ordained" (276). The sentences that immediately follow are instructive: they are emblematic of the fraught, contradictory efforts to inaugurate politics by holding sociality and biology apart. Rubin continues:

> This [social and historical constitution of sexuality] does not mean the biological capacities are not prerequisites for human sexuality. It does mean that human sexuality is not comprehensible in purely biological terms. Human organisms with human brains are necessary for human cultures, but no examination of the body or its parts can explain the nature and variety of human social systems. The belly's hunger gives no clues as to the complexities of cuisine. The body, the brain, the genitalia, and the capacity for language are all necessary for human sexuality. But they do not determine its content, its experiences, or its institutional forms. (276)

In the decades after this declaration, Rubin's political gesture (for the social and away from the biological) became second nature to feminist critique, and the act of peeling biological influence away from social principles became critically habitual. Indeed, without such action, it has often been difficult to see how any argument can lay claim to being feminist or, more broadly, political (Kipnis 2006).

In a surprising number of contemporary feminist texts that have very little, or nothing at all, to do with biology, one of their core conceptual commitments is a repudiation of biological explanation. An antibiological gesture is often the ignition that starts the theoretical engine. Take, for example, Janet Halley's (2006) *Split Decisions*. This text makes a sophisticated intervention into feminist theory, and Halley explicitly states her debt to Rubin's work: "The Traffic in Women," Halley notes, is "the locus classicus of the crucial feminist idea—I rely heavily on it in this book, and so does everyone in this lineage from here on out—that sex and gender are distinguishable. Rubin powerfully demonstrated that the distinction would give feminism a remarkable new range of explanatory powers" (114–115). At the beginning of *Split Decisions*, Halley provides a definition of her key terms: sex, gender, sexual orientation, sexuality. What she means by sex is "penis or vagina, testicles or ovaries, testosterone or estrogen and so forth" (24). She calls this "sex1," to differentiate it from "sex2," by which she means fucking. Sex1 is tightly defined around discrete biological units: organs and chemicals. In contrast to this, Halley defines gender as "everything else" (24) that differentiates men and women: it is a "whole system of social meaning" (24). Following in the tradition set down in "The Traffic in Women" and consolidated in "Thinking Sex," Halley's definition of gender is significantly more capacious than her definition of sex1. Gender is a sizable, intricate semiotic formation; sex1 is narrow and inert and immaterial to the politics at hand. Importantly, Halley does not return to ponder the nature of these biological monads (penis, vagina, testicles, ovaries, testosterone, estrogen) that lie mutely at the beginning of her analysis. While her arguments about sex2 (and its quarrels with gender) are not simplistically derived from, or reducible to, this antiorganic gesture, there is no question that her politics have been rendered legible and legitimate in part by that gesture.

The importance of Rubin's work, then, is not that she single-handedly authored feminist antibiologism, but rather that she was able to so lucidly articulate it and noiselessly embed it within larger, more urgent, foundational arguments about gender and sexuality. In concert with an already existing set of political expectations about biology, "The Traffic in Women" and "Thinking Sex" were important loci of the crucial feminist idea that biology and politics are disjunctive. While the political and theoretical questions raised in "Thinking Sex" are still being vigorously contested, the potent rhetorical gesture that, in part, made these arguments viable (i.e., "no examination of the body or its parts can explain the nature and variety of human social systems" [Rubin 1984, 276]) has been less closely examined. Consequently, many feminist theories still rely on this core contradiction: biology is both a prerequisite and politically irrelevant. It is peripheral to our political concerns, yet it bears down on them dangerously.

In recent years, there has been some restlessness about the need to reject biology. There is growing feeling that the antibiologism on which feminism cut its teeth has now become politically and intellectually restrictive. It is not just feminists working in science studies or the history and philosophy of science who feel constrained by the antibiologism in feminist theory; there is also a broader sense that feminist theory would be made stronger (for all manner of disciplinary projects) by closer engagement with biological detail. Laura Kipnis, for example, in her comments marking the thirtieth anniversary of the publication of "The Traffic in Women," voices her disquiet about how anatomy has come to be regarded in feminism: "Rereading the essay made me reflect that the scope of its influence—echoed in a range of feminist work that followed—has made it rather unacceptable to interrogate truisms such as 'the body is a social construct'" (437). There are certain kinds of anatomical data (about bodily pain, for example) that Kipnis feels have been dismissed under the rubric of social constructionism, and she wonders if we might be able to talk about these biological experiences more candidly, less suspiciously. Yet even as Kipnis frets about the restrictive uses of anatomy in feminist politics, she repeats a conventional presupposition about biology. She still sees biology as inflexible stuff: "The problem with bringing up biology is that you're taken to somehow endorse it. . . . Please understand that I don't endorse these anatomical facts, I'm just stuck with them" (435–436). In both Rubin and Kipnis, there is an anxiety about biology's power to determine form and control politics. Rubin wants to push biology away, Kipnis wants to draw it closer, but neither has yet displaced the shared phantasy that biological matter is sovereign, intransigent, bullying. Is there not a shared belief in Rubin and Kipnis (and more widely) that to engage with biology is to find ourselves stuck?

This book does endorse biology. It vouches for the capacity of biological substance to forge complex alliances and diverse forms. It advocates for a biology that is nonconsilient. It disputes the grandiose notion (increasingly found in science-humanities scholarship) that biology can conclusively resolve questions of psyche or politics or sociality, and it rejects the creed that biological data bring interpretative methodologies to an end (see the conclusion). Instead it seeks out systems of biological overdetermination. In particular, this chapter examines how feminist theory got itself trapped in relation to biology. If, as I argue, there is no intrinsic orthodoxy to biological matter (if it can be as perverse and wayward as any social, textual, cultural, affective, economic, historical, or philosophical arrangement), why have we so readily joined with conventional biologism to think of biology as predetermined matter? What conceptual payoff (what secondary gain) have we received for this? And how easy will it be to do otherwise? I begin by working through a small section of "Thinking Sex" to map out some of the conceptual and political effects of Rubin's aversion to biological explanation. In particular, I am interested in how the belly figures in her attempts to forge new directions for feminist theory. It is the belly that will be central to my Kleinian interest in biological phantasy in the latter part of the chapter.

The Traffic in Biology

Rubin's (1984) claim that "human sexuality is not comprehensible in purely biological terms" (276) is, I think, uncontentious if we keep the focus on the word *purely*. It is true enough that sexuality is not comprehensible in purely biological terms. But then again, nothing is comprehensible in purely biological terms—especially not biology itself. The work of feminist theorists of biology like Anne Fausto-Sterling, Evelyn Fox Keller, and Donna Haraway has been able to show how the gene, or the neuron, or the hormone is from the beginning a biologically impure object. There are no entities or events, they argue, that can legitimately lay claim to being biological and not also cultural or economic or psychological or historical.[2]

Let's take tryptophan as an example. Tryptophan is an essential amino acid and one of the building blocks of the neurotransmitter serotonin. By current neuroscientific reckoning, serotonin is one of the chemicals that modulates mood. A reliable supply of tryptophan, then, is one of the things that is necessary to maintain consistent serotonin levels in the human body, and by extension stable and buoyant mood. Should we infer from these theories that tryptophan can be understood in purely biological terms? Or that the serotonin hypothesis about depression is an exclusively biological claim? The answer to both of these questions is no: tryptophan and serotonin are fabricated by extensive intrabiological and extrabiological traffic. Because tryptophan is an essential amino acid, the human body cannot make it from scratch: it has to be obtained from outside the body. It has to be eaten. Bananas, milk, eggs, lentils, nuts, soybeans, tuna, and rice are all high in tryptophan. A diet dramatically low in tryptophan will leave that individual or that group unable to generate adequate amounts of serotonin. When entwined with other events (personal grief, cultural trauma, transgenerational poverty), low levels of serotonin may have significant, negative effects on individual and group well-being.

It would be something of a miscalculation, then, to see tryptophan as a purely biological object. The very efficacy of tryptophan as a chemical of mood relies on processes of exchange across bodily borders, within cultures, and between animate and inanimate objects. Once tryptophan is ingested, the processes by which serotonin is synthesized are no less complex (more of which in chapter 4). Serotonin levels in the human nervous system are intimately tied to tryptophan and through tryptophan to the belly and to cuisine. In this important sense, serotonin is a biological object impurely constituted through relations to other entities and events, each of which is also impurely and relationally fabricated. The biology of depression, or (to return to Rubin's example) the biology of sexuality, is always doctored. One corollary of this argument—and this has received much less feminist attention than it deserves—is that there are no sexual identities or modes of embodiment or cuisines that can legitimately claim to being cultural but not also biological. Our social objects and structures crystallize in systems of mutuality that include, among other things, neurons and hormones and genes. Likewise, our neurons and hormones and genes crystallize as biological entities in systems of mutuality that include, among other things, social objects and structures. All worlds are alloyed; no object is purebred.

What Rubin seems to be getting at, however, is not so much an argument about what kind of biology (pure/impure) there could be in our theories of sexuality. Rather, at this moment, she is consolidating the notion that in feminist theory there should be no biology at all. This political principle was so widely shared that in 1984 it needed little more than a paragraph or two to be elucidated. Rubin confidently claims that biological substrata (our brains, in this instance) do not determine the content, experience, or institutional form of sexuality. Crucially, she makes this claim without consulting empirical evidence. She does not examine a neurological theory of sexuality; the reader of "Thinking Sex" does not know what the neuroscientific data might disclose about the nature and variety of human neurological systems. This neglect of empirical evidence is a curious turn for Rubin. It violates her own strong commitment, beautifully articulated in the 1994 interview with Butler, that data are the lifeblood of robust feminist research:

> There needs to be a discussion of what exactly is meant, these days, by "theory," and what counts as "theory." I would like to see a less dismissive attitude toward empirical work. There is a disturbing trend to treat with condescension or contempt any work that bothers to wrestle with data. . . . It is a big mistake to decide that since data are imperfect, it is better to avoid the challenges of dealing with data altogether. I am appalled at a developing attitude that seems to think that having no data is better than having any data, or that dealing with data is an inferior and discrediting activity. A lack of solid, well-researched, careful descriptive work will eventually impoverish feminism, and gay and lesbian studies, as much as a lack of rigorous conceptual scrutiny will. (Rubin in Butler 1994, 92)

While feminists have engaged carefully with, say, ethnographic data or sociological data or historical data to build new theories of gender and sexuality, we have been less enthusiastic about data from the natural sciences. In relation to that kind of data we have been almost uniformly suspicious. To return to *Split Decisions*, Halley (2006) uses the words *purported* and *supposedly* to describe conventional (dimorphic) theories of biological sex: "supposedly irreducible fact" and "purported bodily differences" (24). The disdain she has for narrow, morally policed definitions of gender is also mobilized in relation to biology. But where her scorn for gender is expansive (it occupies the entire book and utilizes considerable intellectual energy to very effective political ends), her rejection of biology is massively abbreviated. Short and sharp, the words *purported* and *supposedly* stand in for a shared ground of feminist skepticism about the value of biological data and theories.

It is this twin movement that interests me most: the expansion of theoretically astute feminist argumentation, on the one hand, and the massive contraction of interest in biological substrata, on the other. Here is my suggestion: feminist theory has presumed a kind of biology—a biology that is largely static and analytically useless—as one way of securing its critical sophistication. It is the second part of this claim that carries the real punch. It is not simply that feminist theory has often misread biology (that we have misunderstood it or ignored it, both of which might be fairly trivial events). I have a stronger claim to make: these misreadings and repudiations of biology have had the particular effect of making feminism smart. These misreadings and repudiations have been very profitable: they have helped build our theories and affirm our politics. This means that our theoretical innovations (like everyone else's theoretical innovations) have at their heart an unacknowledged and effective repudiation. My goal, then, is not to read Rubin or Halley as having fallen into error.

The refutation of bigoted biological theories of gender and sexuality has been, and remains, vital. Neither am I campaigning for a feminism that would refuse to make such a repudiation, as if good intentions and prudence would wash the difficulty away. The situation is more complex than that: the rejection of biology has made us who we are, it is spliced into the DNA of feminist theory. This repudiation can't be wished away, and even critiques of this repudiation (such as my own) inevitably carry within them the traces of a desire to be free of biological constraint. This tangled indebtedness to antibiologism isn't easily solved, especially not by a simple anti-antibiologism (the uncritical embrace of neuroscience, for example). It seems to me that as feminist theory comes to engage more closely with biological data, this must also be a task of significant, gradual self-transmogrification. Where that transformation leads is not clear—and for me this is one of the things that makes an engagement with biology so compelling.

One way to work this problematic is to examine how the conventions of antibiologism got laid down inside feminist theory and how they continue to influence what we think our politics should be. To that end, let me continue to follow Rubin's argument. Her antibiologism finds one of its clearest articulations in her declaration that "the belly's hunger gives no clues as to the complexities of cuisine" (Rubin 1984, 276). What I presume Rubin means here is that physiological data about hunger will tell us nothing about the production and consumption of food. Contractions of the stomach walls, changes in blood sugar, liver metabolism, hormonal cascades, nervous activity: these things give us no sign, no evidence, no insights into the rituals of eating or the histories of cooking. She makes the same claim in 1975 in "The Traffic in Women," when she notes (following Claude Lévi-Strauss) that "hunger is hunger but what counts as food is culturally determined and obtained" (Rubin 1975, 165).

Two effects (at least two) ripple out from this kind of statement. First, the trade between nature and culture is greatly diminished, if not halted altogether. Biological systems and cultural systems become autonomous, each operating according to its own internal logic: one gives no clue about the other. The conceptual foundation for this division was formalized in "The Traffic in Women," where Rubin defines the sex/gender system as a fairly simple nature/culture interaction: "the set of arrangements by which a society transforms biological sexuality into products of human activity" (159). This means that in Rubin's sex/gender system biology is passive substrate ("raw material" [165]) that culture animates. Kinship—the core conceptual device in that essay—is similarly described as "the imposition of cultural organization upon the facts of biological procreation" (170–171), and psychoanalysis is said to explain "the transformation of the biological sexuality of individuals as they are enculturated" (189). This notion of torpid materiality would be brought into question by feminist theorists of embodiment in the 1980s and 1990s. But those theories often entrenched our suspicions that biology is treacherous material, even as they breathed life into "the body" (Wilson 2004). Thus Kipnis's sense of being stuck. Kipnis rightly understands that the separation of biology and politics is unsustainable; yet feminist theory, in its canonical and most accomplished forms, tends to reinforce rather than reconfigure the segregation of these fields. We need more ways of reading for how clues about biology might also be clues about politics, and vice versa.

Second, Rubin's separation of belly and cuisine is not analytically neutral. This gesture attributes complexity to cultural production (to cuisine), and it attributes simple-mindedness to biological events (to hunger). One effect of this is to make a preference for the former over the latter all but inevitable; and this kind of preference meshes with a series of other distinctions that are routine in feminist theory. Most of us trade in the difference between things that are complex and things that are reductive, things that are political and things that are conventional, things that are theoretical and things that are empirical. Indeed, is it not within such systems of distinction that feminist theory has forged much of its analytic sophistication? Joseph Litvak (1997) has offered a compelling reading of this tendency to oppose the sophisticated and the vulgarly literal. He argues that while establishing a position of sophistication often seems to entail keeping a distance between oneself and the literal (especially the vulgarity of alimentary processes), sophistication is in fact intimately connected to basic bodily appetites. Academic theories, in particular, are prone to wanting to keep their distance from rudimentary bodily events: "In talking about sophistication, one needs to keep all these terms—the culinary, the erotic, the linguistic, the economic—in play; [because there is] a certain tendency toward abstraction in academic commentary, the [terms] that risk dropping out first are the more literal or corporeal ones rather than the more symbolic or social ones" (8). Rubin's work could hardly be said to lack corporeal or erotic interests. But it does lack interest in bodily matters (hunger, mastication) deemed less cultured.

In 1975, Rubin argues—in a somewhat utopian vein—that "a full-bodied analysis of women . . . must take everything into account" (209). However, the transition from 1975 to 1984 actually sees her taking less biology into account. Without question, there is a proliferation of sexualities in 1984 (transsexuals, fetishists, sadomasochists, transvestites, pederasts, and prostitutes: a gathering of the clans of perversity), but there has been a dramatic narrowing of both the biological character of politics and the political nature of biology. A surprising amount of biology is at play in the 1975 argument. By 1984, however, little is left: the role of biology has been cheapened to being simply the marker of the place away from which analytic elegance and political acuity have evolved. Since then this kind of antibiologism has become both routine and almost entirely silent. For example, none of the papers in the 2011 collection in the journal GLQ that commemorates the twenty-fifth anniversary of "Thinking Sex" engages with the biological arguments of Rubin's paper. Biology appears only once, in a review of the conference that begat the collection: "In 'Thinking Sex,' Rubin referenced race by way of analogy to sex. Like race, she proposed, sexuality is given life and meaning by culture and history, not by biology" (Kunzel 2011, 161). The routine way in which biology is dismissed, and the fact that this core argument (race is like sexuality, but neither is like biology) is not elaborated on or interrogated in any way, speaks to the holding power of antibiologism for so much contemporary feminist theory.

It seems clear enough, then, that feminist theory has developed in concert with antibiologism and that the finesse of many feminist theories draws, in a nontrivial way, on the presumption that biology is peripheral to our political interests. At the same time (and seemingly in contradiction of these habits), feminist theory often treats biology as a threat. Biological data are thought to imperil our political and conceptual progress: "It is impossible to think with any clarity about the politics of race or gender," Rubin (1984, 277) argues, "as long as these are thought of as biological entities. . . . Similarly, sexuality is impervious to political analysis as long as it is primarily perceived as a biological phenomenon." For Rubin, biology is an obstacle to politics: it is a tyrant to be overthrown, it is the constraint against which we struggle to imagine other configurations of the body and sexuality. At these moments, she finds herself ensnared in something like the repressive hypothesis: "We will not be able to free ourselves from it except at a considerable cost: nothing less than the transgression of laws, a lifting of prohibitions, and irruption of speech, a reinstating of pleasure within reality, and a whole new economy . . . will be required" (Foucault 1978, 5). Perhaps there is a strain of political sentiment that binds antibiologism and the repressive hypothesis together, or at least increases the likelihood that they will occur together, even in muted form, in feminist arguments: in both cases there is a commitment to a juridical notion of power that acts as a sovereign, bullying source of authority, and in both cases there is a political demand that we rally, for the good, against this pernicious influence.

Rubin's argument in "Thinking Sex" is exemplary in this regard, and I wonder if her antibiologism doesn't draw some of its strength from her fidelity to a repressive account of sexuality. While she aligns herself explicitly with Foucault's first volume of *The History of Sexuality*, the first sections of "Thinking Sex" disregard Foucault's surprising assertion that sexuality has been governed by proliferation rather than prohibition. Rubin's opening pages reaffirm a version of the repressive hypothesis by documenting how deviant sexualities have been subjugated by proscriptive forces that date from the Victorian era: "The consequences of these great nineteenth-century moral paroxysms are still with us. They have left a deep imprint on attitudes about sex" (268); "for the last six years, the United States and Canada have undergone an extensive sexual repression" (270); "new erotic communities, political alliances, and analyses have been developed in the midst of the repression" (275); "sexual speech is forced into reticence, euphemism and indirection" (289).

Rubin's essay organizes its politics around a juridical (moral) power that subjugates sex. Indeed, what Rubin draws from Foucault is less his critique of repressive power than his alleged stance against biological accounts of sexuality. Foucault is important, she argues, because he is part of a new scholarly tradition that imagines sex other than through biology: "Foucault criticizes the traditional understanding of sexuality as a natural libido yearning to break free of social constraint. He argues that desires are not preexisting biological entities, but rather, that they are constituted in the course of historically specific social practices" (276). Leaving aside whether or not Foucault makes such a strong distinction between biology on the one hand and social practices on the other, it is clear that Rubin has aligned biological theories of sexuality with repressive politics. To think biologically is to think coercively. Rubin's antibiologism and her commitment to thinking of power as an oppressive force fit into the same conceptual schema. Her antibiologism and her repressive model of sexuality seem to find succor in each other: together they entrench the belief that feminist politics should be predicated on the overthrow of juridical-biological power.

One of the things I would like to show in the latter chapters of this book is that close attention to biological theories and biological data undermines this notion of the organic as a juridical force. Biological authority does not emerge from a singular origin (the brain, say) and radiate out from there, in a domineering fashion, to control the rest of the body and then the behavioral and social worlds. Rather, biological data often look much more like they are describing networks of affinity where organic entities build their natural capacities through mutual, and often asymmetrical, relations with each other (I will elaborate on this in the next chapter in a discussion of "amphimixis"). Much of the recent turn to neuroscience in the humanities and social sciences has been so unappealing, it seems to me, because it fails to question the conventional notion of biological authority; it simply seeks to harness a juridical biology to humanistic ends. The ease with which this kind of orthodox scholarship has found circulation in the academy shows that the notion of juridical biology still holds considerable appeal, especially for those trained outside these fields of expertise. This general logic of a magisterial biology has been extremely hard to displace, and it has been difficult to find ways of engaging with biological data that are not governed by the demand that one is either for it or against it. If feminist theory is to continue to make trouble, it will need to form intimate and unruly alliances with biological data. We need these kind of alliances with biology not just in relation to depression; more generally they help unsettle the political certainties of what we think we stand for, what we think we stand against, and where we stand when we make political gestures.

At the end of "The Traffic in Women," Rubin muses about a future without anatomy: "I personally feel that the feminist movement must dream of more than the elimination of the oppression of women. It must dream of the elimination of obligatory sexualities and sex roles. The dream I find most compelling is one of an androgynous and genderless (though not sexless) society, in which one's sexual anatomy is irrelevant to who one is, what one does, and with whom one makes love" (204). I will explore this peculiar (and impossible) wish to be rid of anatomy in depth in the second chapter in relation to the work of Sigmund Freud and Sándor Ferenczi. In preparation for that longer discussion I turn now to the work of Melanie Klein. Through her ideas about the hungry belly of an infant (and the rudimentary phantasy and biology that the belly is thought to encapsulate) I will begin an alternative narrative about the important work that anatomy can do for feminist theory.

The Biology of Phantasy

One response to the charge that feminism got smart by refusing biology is that we might take more interest in the debris that sophisticated feminist theory leaves in its wake: the concrete, the literal, and the reductive. What value might these remains have for doing theory otherwise? It seems to me that there are few writers better suited for this task than Melanie Klein: who else immerses us in the concrete, the literal, and the reductive with greater effect than Mrs. Klein and her followers?[3]

One of the things that will strike critically schooled readers when they first encounter Klein is that her work seems hardly theoretical at all. In comparison to the elegant conceptual machinery that is Freud's writing, Klein's work is much more phenomenologically concrete in its examples and much less concerned with theoretical finesse. Because Klein does most of her best work with simple analytic pairs (love and hate, good and bad, introjection and projection, envy and reparation, part and whole, phantasy and reality), a theoretically proficient reader might be alarmed by the interpretations of clinical material that form the core of Klein's work. For example, Klein reports that Patient X developed diarrhea (which he mistakenly thought was mixed with blood) when his tapeworm phantasies were being explored in analysis. Her interpretation of this symptom feels compacted and blunt: "This frightened him very much; he felt it as a confirmation of dangerous processes going on inside him. This feeling was founded on phantasies in which he attacked his bad united parents in his insides with poisonous excreta. The diarrhoea meant to him poisonous excreta as well as the bad penis of his father. The blood which he thought was in his faeces represented me" (M. Klein 1935/1975, 273). To be engaged with Freud is to be in love with theory. To be engaged with Klein is to be seduced by data that seem too raw (and, as we will see, too biological).[4]

The tone of Klein's clinical and theoretical interpretations was one of the concerns voiced during the "Controversial Discussions," when members of the British Psycho-Analytical Society met (1941–1945) to discuss the details of Klein's work and its impact on classical Freudianism (King and Steiner 1991). Marjorie Brierley was a strong critic in this regard. She argued that Klein confuses the subjective voice with the scientific voice, and the phantastic with the physiological:

A failure to distinguish sharply between the *patient's* language and *scientific* language is not in the least a merely linguistic mistake but may result in capital confusions or errors in our view of real events. This vital distinction between subjective and scientific descriptions of scientific events may also be illustrated by a bodily pain due to a physical stimulus. Thus, if a man with an over-acid stomach eats a sour green apple, he will probably experience pain. He may say about that pain: "That apple is burning a hole in my stomach." This, in effect, is a phantasy; the processes set up in his stomach feel like that. The things that are actually happening in his stomach will be more accurately inferred and described by the physiologist. (Brierley 1942, 108–109)

It seems to go without saying, for Brierley, that phantasy and physiology are separate ontological realms that require distinct lexical choices. Klein's "mistake" is that she has been unable to maintain a sharp distinction between the physiology of the stomach and the phantasy of the stomach, and between the language used to describe each. This tangle, Brierley argues, makes it difficult to build a reliable and respectable psychoanalytic science. Indeed, part of the exasperation expressed with Klein during the controversial discussions derives not so much from her lack of conceptual precision but from a recognition that in her work there is no ontological clarity (no bright dividing line) between minded states and biological processes, and that this stance threatens to bring psychoanalysis into disrepute (J. Rose 1993). How can any science work without clearly delimited variables?

In a similar vein, Brierley is also concerned that Klein has been unable to partition the phantasies of a child from the theorizing of an adult; "she is so keenly alive to the child's actual beliefs that she sometimes gives the impression of explaining her theory in terms of these beliefs" (109). For this reason, Brierley argues, Klein's theory threatens to become animistic (attributing aliveness to inanimate objects). In a move that Brierley might disparage as conceptually infantile, Klein lets the animate and inanimate worlds interbreed.[5] I am arguing that these confusions (if that's what they are) between things that are alive and things that are not, between archaic states and adult theories, between incorporation and digestion might be a theoretical exemplar for feminist engagements with biological data. If we start with the presumption that mind and gut are keenly alive to each other rather than disengaged, perhaps our political intuitions (for cuisine; against the belly) can be rescripted. In particular, perhaps we can move away from a politics primarily informed by the rhetoric of domination (biology!) and rebellion (culture!) and look for theories that exploit the logic of imbrication.[6]

Eve Kosofsky Sedgwick (2007) has identified this potential for Klein to move us athwart our usual critical and political practices. She calls Klein's work "chunky," and she means this as a compliment. She compares Klein's corpus to the kind of oversized doll she (Sedgwick) had once demanded as a young child: "I needed . . . something with decent-scale, plastic, resiliently articulated parts that I could manipulate freely and safely . . . where the individual moving parts aren't too complex or delicate for active daily use" (627–628). It is this aspect of Klein's work that provides Sedgwick with another route for intellectual engagement: "As someone whose education has proceeded through Straussian and deconstructive, as well as psychoanalytic, itineraries where vast chains of interpretive inference may be precariously balanced on the tiniest details or differentials, I feel enabled by the way that even abstruse Kleinian work remains so susceptible to a gut check" (628). Sedgwick notes that in comparison to the Freudian landscape, which is populated with representations, the Kleinian landscape is populated with things (objects). She calls this orientation in Klein a "literal-minded animism" (629), which again (contra Brierley) she means as a compliment.

The advantages of this literal-minded animism are nowhere better played out than in the Kleinian account of phantasy in the very young infant. Phantasy at this age is coterminous with physiological states. Specifically, it is events in the infant belly that are a central part of Klein's theory of mind. She recognizes, of course, that the infant uses its whole body to take in the world: "The child breathes in, takes in through his eyes, his ears, through touch and so on" (M. Klein 1936/1975, 291). Nonetheless, the gut is a privileged site for the operations of infant phantasy. Where Freud's theory of mind speaks to the importance of the surfaces and openings of the body in early infancy (the mouth and the anus), Klein digs down into the center of the body, into the stomach: "The first gratification which the child derives from the external world is the satisfaction in being fed. . . . This gratification is an essential part of the child's sexuality, and is indeed its initial expression. Pleasure is experienced also when the warm stream of milk runs down the throat and fills the stomach" (290).[7] Klein disputes the idea that this experience is autoerotic or narcissistic in the sense that the infant takes itself as its own object, oblivious to the world. The infant is certainly internally directed, for Klein, but not in a solipsistic way. Rather, the infant is in intensive relations to internal objects—to parts of the world, parts of its body, parts of other people that have been taken in through the gut. Right from the beginning, other things are a core part of me. Right from the beginning, I am impurely, relationally, enterically constituted.

In the controversial discussions, Anna Freud clearly outlines the theoretical and clinical distinction between herself and Klein on this point: "For Mrs. Klein object relationship begins with, or soon after, birth, whereas I consider that there is a narcissistic and auto-erotic phase of several months duration, which precedes what we call object relations in the proper sense, even though the beginnings of object relations are slowly building up during this initial stage" (in King and Steiner 1991, 418–419). Of the many things that were being negotiated in these discussions one of the most important is when object relations begin. For Klein this is "almost from birth" (M. Klein 1936/1975, 290); for Anna Freud it is several months later. I am less interested in adjudicating on that particular issue than I am in its implications for theorizing biology and mind. It seems to me that debates about when object relations begin are implicitly arguments about the nature of biology: What status do we give to the biological processes that lie, allegedly, before object relations (in utero, at birth, at six months of age)? What is the biological nature of the drive before it makes contact with the external world? The search for a starting point to object relationality presumes that there are biological processes (hunger, for example) that have not yet been brought under the sway of phantasy. These debates usually presume that psychic action is a secondary elaboration of prior (purely?) biological stimuli. Here I would like to press on Kleinian notions of phantasy and the belly in order to dissolve the idea that there is a developmental point before which biology is not minded.

The infant's mind emerges out of phantasmatic relations to incorporated shards of other people, parts of the body and the world. Are these objects good or bad, do they hurt me or soothe me, do I want to take them in or spit them out? Attachments and detachments that emerge at a later date (e.g., depressive moods) have as their prototype these early relations to the things the infant has swallowed. The hunger pangs of the infant's stomach are crucial to the structure of mind and the capacity to be attached: they are among the first stimuli the newborn will negotiate, and the enormous psychic force of an empty stomach or a full stomach is the reason Klein and her followers claim that phantasy is present almost from birth. The gnawing of hunger inside the body will be felt as a persecutory object that is inside me; not inside me in an abstract kind of way (an idea in my head), but a destructiveness that is literally inside my belly. The belly is home, then, to both good and bad objects in concrete, animistic ways. Put another way, we could say that the belly is psychically alive to the infant. The first mind we have is a stomach-mind.

Susan Isaacs—one of Klein's strongest advocates—claims that the difference between a phantasy and a physiological process is moot for the very young infant.[8] Quoting Clifford Scott (a fellow Kleinian), Isaacs (1948) notes that "the adult way of regarding the body and the mind as two separate sorts of experience can certainly not hold true of the infant's world. It is easier for adults to observe the actual sucking than to remember or understand what the experience of sucking is to the infant, for whom there is no dichotomy of body and mind, but a single, undifferentiated experience of sucking and phantasying" (86). Eventually, the infant begins to distinguish between sensation and feeling, phantasy and reality, inside and out. The composite existence of sucking-sensing-feeling-phantasying becomes "gradually differentiated into its various aspects of experience: bodily movement, sensations, imaginings, knowings, and so on and so forth" (86), although these events are never fully autonomous from each other at any time in life. Knowing is also moving; sensing is also imagining. Here we seem to be close to what could be called the biological unconscious: substrata that are able to act organically and psychologically at the same moment (see chapter 2). At this point, object relations are also organ relations; and later psychological states will always be endowed with that archaic capacity to suck-sense-feel-phantasize. Isaacs's point is not that raw hunger will be elaborated, secondarily, in the infant's mind, generating phantasies ("That apple is burning a hole in my stomach"). The implication of her position is that the biology of hunger is already and always a minded event: the contractions of the stomach walls, changes in blood sugar, liver metabolism (events that I have suggested Rubin disregards) are phantastically alive—from birth, before birth, in prehistory. Infant minds emerge from an engagement with this unconscious biological mentation.

In an otherwise astute reading of the value of Klein's and Isaacs's work for feminist theory, Jacqueline Rose (1993) misses the opportunity to think biology otherwise. Indeed, one of her initial gestures is to eject biology from the scene of rereading Klein. Just as Halley clears away sex1 in order to think about gender, Rose notes: "The *first thing* that becomes clear is that the concept of the death instinct or impulse is in no sense a biologistic concept in the work of Klein" (148, italics added). By attributing such biologism to Anna Freud, Rose moves over the nature of biology in Klein too quickly. Defending Kleinianism against critics who have read her work as biologically reductionist, Rose argues that Kleinian negativity is not a "biological pre-given" (169) but rather the phantastic subversion of biology. Here I want to relinquish this distinction between a biological pregiven, on the one hand, and the subversion of biology (secondarily, by phantasy) on the other. I want to use Klein (and, in the next chapter, Ferenczi) to suggest that what is "pregiven" in biology is phantasy. There is no need to defend against biological reductionism in Klein, because biology for Klein is already and always (in its most reduced form, if you like) a phantasmatic substance.

This Kleinian position on biology and phantasy draws much of its conceptual strength from the phylogenetic hypotheses that circulated widely in the early years of psychoanalysis. For many analysts at this time (Freudian, Kleinian, Jungian) some aspect of phantasy is deemed to be inherent in bodily impulses that have been passed down from earlier, archaic times.[9] For example, in addressing the objection from her peers that unconscious phantasies of aggression (e.g., to tear my mother to bits) require that the infant has explicit knowledge that tearing to bits means killing or direct experience of the consequences of such aggression, Isaacs replies,

> Such a view is really absurd. It overlooks the phylogenetic source of knowledge, the fact that such knowledge is *inherent* in bodily impulses as a vehicle of instinct, in the *aim* of instinct. When Freud says that the aim of oral love is "incorporating and devouring, with abolition of any separate existence on the part of the object," does he mean that the infant has seen objects eaten up and destroyed, and

then comes to the conclusion that he can do this too, and so wants to do it? No! He means that this aim, this relation to the object, is inherent in the character and direction of the impulse itself, and its related affects. (In King and Steiner 1991, 451–452)

At these moments, when there is no clarity between individual mind and species inheritance, between biological impulses, sensations, affects, and phantasies, Kleinianism is at the very edges of scientific respectability. Most often Kleinians have drawn back from this precipice and have secured for themselves more orthodox epistemological foundations. In particular, the call of a developmental narrative (the need to explain how the infant becomes, psychologically, a child and then an adult) usually has the effect of rendering the biology of sucking-sensing-feeling-phantasying more tame, and this generates a dishearteningly conventional theory of mind's relation to bodily substrate.

It has been argued, for example, that at a particular tipping point in the infant's development the stomach-mind cedes ground to the capacity for abstraction. Robert Hinshelwood, the original author of the respected *Dictionary of Kleinian Thought*, calls this juncture, hyperbolically, "a glittering moment in the history of each individual" (Hinshelwood 1991, 352). At this point, he notes, something crucial changes in mind-body relations: "Phantasies *about* the bodily contents stand for the actual primary bodily sensations" (38). That is, metasomatic capacities (the ability to think about the body, to represent the body, to stand at an affective distance from the body) become part of the mind alongside rudimentary somatic states. The body no longer has to act out what cannot be tolerated: it can be cognized and spoken and reformulated. Pleasures, too, can be transmuted from primal sensations to socially viable events. As Hinshelwood continues, however, these various psychosomatic states, hitherto cohabiting in the mind, are organized into a conventional developmental hierarchy in which higher symbolic and cognitive expertise arises out of, and leaves behind, a primordial soup: "Subsequently, the infant emerges into the social world of symbols in which phantasies are composed of non-bodily and non-material objects. The movement from a concretely felt experience of an object, constructed in unconscious phantasy, to a non-physical symbolic object is a major developmental step" (38). What tends to drop out of this picture, as Litvak might remind us, is the psychic force of concrete bodily states. There is a tendency in Hinshelwood (and others) to proceed as if higher capacities are autonomous from their primitive bedrock and as if the bedrock itself has no phantastic capacities.[10]

As I read Hinshelwood, I am reminded again of Rubin's conviction that the move away from the biological and toward the social epitomizes a more evolved attitude. In Hinshelwood's developmental schema, somatization tends to be read as regressive and infantile; in Rubin's conceptual schema, biological substrate is the very last thing with which you would want your politics implicated. I remain convinced that a theory of stomach-mind moves across the conventionalism of both Hinshelwood and Rubin in important ways. A phantastic theory of biology strikes me as critical to feminists as they struggle to get themselves unstuck in relation to biological determinism, and as they seek ways to politically engage that are not always caught in juridical positions of for/against.

Conclusion

I have been tracking the psychic character of the organic interior as a way of formulating a different feminist engagement with biology: a kind of critical splanchnology (the scientific study of the viscera). It is not clear to me that feminist theories can easily or immediately extricate themselves from the antibiologism that has hitherto underwritten much of their critical sophistication; but they can certainly engage— strongly and enthusiastically—with the sequelae of that antibiologism. Rather than working to avoid, expose, or eliminate biologism (which strikes me as something of a Sisyphean task), perhaps it is possible to turn, with more curiosity, to the debris that our antibiological politics have generated: archaic action, animism, rudimentary mentation. The belly feels like just the right kind of container for such endeavors. *Gray's Anatomy* (Gray 1918) notes that "the shape and position of the stomach are so greatly modified by changes within itself and in the surrounding viscera that no one form can be described as typical" (1161). The belly takes shape both from what has been ingested (from the world), from its internal neighbors (liver, diaphragm, intestines, kidney), and from bodily posture. This is an organ uniquely positioned, anatomically, to contain what is worldly, what is idiosyncratic, and what is visceral, and to show how such divisions are always being broken down, remade, metabolized, circulated, intensified, and excreted. If we also take this stomach to be a Kleinian organ, then conventional divisions of mind and body are similarly entangled, and perhaps the motivations of the patient that Leader describes at the beginning of the chapter ("her next thought was to swallow all the pills together") appear in a new light: we can see that one of the gut's archaic feats is minding, apprehending, caring. The next chapter elucidates one particularly galvanizing theory of these substrata in action (Sándor Ferenczi's notion of a biological unconscious) and uses this to think about what might be at stake in cases where swallowing has become phantastically disordered.

CHAPTER 2

THE BIOLOGICAL

UNCONSCIOUS

The Freudian dictum "anatomy is destiny" has been anathema to feminist politics. Condensing the very worst biological reasoning with a troubled account of gender and sexuality, this Freudian declaration has become emblematic of the kind of biologism that feminist theories decry. This chapter explores what anatomy means in certain modes of Freudianism, and how that kind of anatomy has influenced the biological politics that many feminist theories now implicitly affirm. What if what is wrong with this dictum is not so much the claim for predestination, but the presumption that anatomy is so solidly and immutably composed that it could underpin strict determinism? Is there a way to think about anatomy that might undercut claims for its ability to control politics but without abandoning anatomy all together? Building on the claims of the prior chapter, I will suggest that anatomy enacts the kinds of malleability, heterogeneity, friction, and unpredictability that feminist theories can relish. This chapter seeks out that sort of anatomy in the work of one of Sigmund Freud's most loyal and provocative colleagues: Sándor Ferenczi. With Ferenczi I want to show that anatomy is volatile enough to generate multifaceted and paradoxical destinies. Rather than turning away from anatomy, we could turn toward it more attentively to see what improbable capacities it holds. I start this reading where so much innovative feminist work on the body has begun: with Freud.

In 1893 Freud published a small technical paper in the French journal *Archives de Neurologie*. This paper ("Some Points for a Comparative Study of Organic and Hysterical Motor Paralyses") had been drafted many years earlier, probably in 1888, following his sabbatical at the Salpêtrière under Jean-Martin Charcot. For reasons that are not entirely clear, the paper was not published for five years; by then Freud had formed a working alliance with Joseph Breuer and together they had published a preliminary communication on the treatment of hysteria. The first three sections of the 1893 paper are primarily neurological in orientation; the fourth and final section must have been written at a later date under the influence of his work with Breuer, and it is this section that contains an important conceptual argument about hysteria and anatomy. The 1893 paper incorporates two kinds of Freudian approach to the body—one neurological, one psychological. More specifically, the paper documents Freud's transition from one mode of analysis (neurological) to another (psychological). What was the nature of that conceptual transition? And why does it matter to feminists interested in biology?

Hysterical and organic paralyses, Freud (1893a) argues, present clinically in quite different forms. While hysterical paralyses are notorious for their capacity to mimic organic paralyses, in fact they differ from organic conditions in important ways: for example, hysterical paralyses are excessively intense, and they are more precisely delimited in their effects than organic conditions. That is, one would expect a hysterical paralysis to be more thoroughgoing than an organic paralysis, and it would be more strictly demarcated in the body. For example, just the hand, the thigh, or a shoulder would be affected, whereas organic paralyses tend to implicate adjacent parts of the body. Furthermore, hysterical paralyses disobey fundamental rules that govern organic afflictions. In hysterical paralysis, for example, proximal parts of the body, such as the shoulder or the thigh, may be more paralyzed than distal parts, such as the hand or foot. This never occurs in an organic paralysis.

How is it that hysterical paralyses can closely mimic organic paralyses, yet diverge from them in significant ways? More curiously, how is it that hysterical paralyses are able to accomplish biological transformations beyond the capacity of organic pathologies? To solve this clinical puzzle, Freud (1893a) enacted a conceptual distinction that would be very influential on feminist theories of embodiment: he detached the hysterical body from the anatomical body. Organic paralyses, he asserts, are the result of an underlying biological lesion; more precisely, they are governed by "the facts of anatomy—the construction of the nervous system and the distribution of its vessels" (166). Charcot had hypothesized that hysterical paralyses are also the result of a lesion— what he called a dynamic or functional lesion. But Freud disputes this homology. There is no anatomical influence in hysteria, he argues; rather, "the lesion in hysterical paralyses must be completely independent of the anatomy of the nervous system, since *in its paralyses and other manifestations hysteria behaves as though anatomy did not exist or as though it had no knowledge of it*" (169). Hysteria appears ignorant of the anatomy of the body. Hysteria is uninterested in the facts of how muscles, ligaments, nerves, organs, and blood vessels are mapped, how they converge or dissociate, how they connect to distal parts of the body, or how they rely on certain signals or pathways in order to function effectively. Rather, hysteria "takes organs in the ordinary, popular sense of the names they bear: the leg is the leg as far up as its insertion into the hip, the arm is the upper arm as it is visible under the clothing" (169). Hysteria is an alteration of the everyday body (especially as it is understood through tactile and perceptual data); it is an engagement of the body as we know it colloquially—as we imagine, love, or despise it. It is for this reason, Freud argues, that he has never observed—nor will anyone ever observe—a hysterical hemianopsia. Hysteria "has no knowledge of the optic chiasma, and consequently it does not produce hemianopsia" (169).[1]

This early neurological work, in conjunction with his psychotherapeutic treatments of neurotic patients, laid the foundation for Freud's account of conversion hysteria. He claimed that conversion hysteria is the transformation of psychic conflict into somatic symptoms—such as paralysis, pain, numbness, or, most famously in the case of Dora, nervous coughing. The dissociation of somatic symptoms from anatomical constraint is central to this account of conversion—organs, limbs, and nerves are transformed according to a symbolic or cultural logic rather than according to the dictates of anatomy. In the 1893 paper, Freud makes this dissociation explicit; he imagines conversion in terms of ideational mechanisms rather than (and in contradistinction to) biological injury. Asking permission to move from anatomical to ideational ground, Freud argues that "the paralysis of the arm consists in the fact that the conception of the arm cannot enter into association with the other ideas constituting the ego of which the subject's body forms an important part. The lesion would therefore be *the abolition of the associative accessibility of the conception of the arm*. The arm behaves as though it did not exist for the play of associations" (170). The important conceptual point for Freud is that in hysteria the "material substratum" (i.e., cortex) is undamaged, but ideas about the body have undergone some kind of alteration. The idea of the arm, for example, has become associated with a large "quota of affect," and this prevents it from being involved in any associative links with other ideas or organs. It has become ideationally sequestered (lost to consciousness) under the weight of this quota of affect; the arm is now paralyzed. It is liberated from its paralysis only when this affective burden is removed, and the idea of the arm becomes accessible again to "conscious associations and impulses" (171).

This model of hysteria, and Freud's emerging preference for psychogenic explanations over neurological ones, has been enormously influential on feminist accounts of embodiment. The idea that psychic or cultural conflicts could become somatic events was one of the central organizing principles of feminist work on the body in the 1980s and 1990s. This model allowed feminists to think of bodily transformation ideationally and symbolically, without reference to biological constraints; to think about the body as if anatomy did not exist. This chapter argues that the dissociation of ideation and biology is now of limited use for feminist critique. It is not my intent to dispute the power of Freud's account of hysteria or to imply that all feminist rejections of biology can be traced to Freud.[2] Most of all, I do not mean my critique to count as a rejection of Freudian methodologies. On the contrary, I begin with Freud precisely because of the importance of his work for thinking biology dynamically (Wilson 2004). What Freud's approach does not pay attention to, however, is the role played by biology that is not damaged. It has very little to say about the nature of the everyday, minute-by-minute, routine action of biological systems (e.g., surges of biochemicals; metabolic activity; synaptic communication; muscular contractions) that must be part of hysterical symptomology. Put another way, in discounting an etiology based in biological damage, Freud minimizes the role of a broad spectrum of biological substrata in the provocation, maintenance, and treatment of conversion symptoms.

I am interested in interrogating the intellectual corollaries that Freud's model appears to have engendered for the contemporary feminist scene: that the most compelling analytic registers for thinking about the body are symbolic, cultural, ideational, or social rather than biological, and that political or intellectual alliances with the biological sciences are dangerous and retrograde. It is my concern that we have come to be astute about the body while being ignorant about anatomy and that feminism's relations to biological data have tended to be skeptical or indifferent rather than speculative, engaged, fascinated, surprised, enthusiastic, amused, or astonished. The more recent turn to biological, especially neurological, data in feminist and cultural critique often complicates rather than alleviates this dilemma. As I will argue further in the conclusion, there is frequently a credulity in the way that biological data are mobilized in the humanities and social sciences, as if such data reign over interpretative analysis or as if they enforce the final, factual limits of what can be imagined, elucidated, or craved. These new uses of biology are too timid and too obedient, and in the end they are simply the other side of the old feminist repudiations of biology. Taking bulimia as its case in point, this chapter argues that biological and pharmaceutical data are indispensable to feminism's conceptual and political efficacy, but that the use of these data has to be eccentric and unsettling to feminism and biology alike. What anatomy (specifically, the gut) can know, in hysterical and nonhysterical states, is one of the vital lessons to emerge from the eating disorder epidemic, and an analysis of the biological enactments of bulimia can provide a means for thinking anew about the nature of mindbody. In order to provide a conceptual framework for these claims, I will examine one particularly astute response to the classical Freudian theory of bodily conversion—Ferenczi's attempts to inaugurate a biological unconscious.

Organic Thought

In a letter to Ferenczi on July 22, 1914 (Ferenczi and Freud 1996, 6), Freud states his reason for putting off an upcoming summer visit: "I have . . . not sacrificed our usual get-together to comfort but rather to renewed work, for which I can't use comradeship. I also don't work easily together with you in particular. You grasp things differently and for that reason often put a strain on me." Ferenczi's reply the following day affirms that his work can be difficult for Freud:

I have also known for a long time that I "grasp things differently" from you, and that you can pursue my work plans only with considerable strain. . . . Certainly my reason tells me that the manner in which you grasp things is the correct one; still, I can't prevent my fantasy from going its own way (perhaps astray). The result is a mass of ideas which never become actualized. If I had the courage to simply write down my ideas and observations without regard for your method and direction of work, I would be a productive writer, and, in the end, numerous points of contact between your results and mine would still be the result. (Ferenczi and Freud 1996, 8)

The extensive correspondence between Ferenczi and Freud is often painful to read, as Ferenczi (some seventeen years Freud's junior) reels between adoration of Freud, on the one hand, and, on the other, a longing to push psychoanalysis into strange, wild territory and to drive Freud beyond the limits of his tolerance. When Ferenczi did find the courage to write down his ideas in a clinical diary (in 1932, the year before he died), what materialized was an astonishing set of hypotheses, clinical fragments, experiments in technique, and—at the end—a touching and lucid account of his fundamental difficulties with Freud and Freudianism (Ferenczi 1988, 182). By this point, however, their intellectual and professional alliance had been gravely damaged; the strain Ferenczi was putting on Freud and psychoanalysis was too much for everyone to bear.[3]

In these later years Ferenczi was troubled by how to treat trauma analytically. He was concerned that the conventional analytic stance of neutrality and abstinence would be ineffective—or worse—when dealing with traumatized patients. The renewed interest in Ferenczi in the contemporary clinical literature has tended to focus on his work with trauma and his innovations with analytic technique (viz., working with transference-countertransference in a more intersubjective way).[4] However, there is another important point of comparison between Ferenczi and Freud that has been less widely commented on: Ferenczi remained more interested in biological explanation than Freud did. As Freud was elaborating complex psychological (ideational) explanations for conversion, Ferenczi was becoming fascinated with the biological material itself: Can we explain the mechanism of hysterical conversion in biological terms? What does a hysterical conversion tell us, not only about the psyche but also about the character of biological substrate? In the Rat Man case history, Freud comments on the "leap from a mental process to a somatic innervation" that is emblematic of conversion hysteria, but he claims that this leap "can never be fully comprehensible to us" (Freud 1909b, 157). One of Ferenczi's early ambitions was to make the bodily transformations of conversion more intelligible; his efforts in this regard may prove to be uniquely instructive for feminists looking to situate their theories in a more dynamic relation to biological data and theories. Here I sketch out Ferenczi's thoughts on the character of biology—on what it is that the body comes to know in states of extreme psychological distress and how a synthesis of biology and psychoanalysis (what he eventually calls bioanalysis or depth biology) is necessary to understand the character of not just hysterical states but any biological substratum.

Ferenczi opens an important 1919 paper on hysterical conversion with a citation from Nietzsche: "You have travelled the way from worm to human being and much in you is still worm" (Ferenczi 1919, 89). Introducing a writer that Freud famously claimed never to have read, Ferenczi initiates two important provocations to the classical Freudian project: (1) an engagement with psychosis, and (2) an interest in the phylogenesis of the human psyche (the wormlike, regressive nature of the mind). It was differences over the treatment of psychosis and trauma that became such a source of difficulty between the two men in Ferenczi's last years. These differences were not resolved either personally or analytically, and the schism between classical and more relational or intersubjective techniques still structures the practice of psychoanalysis today. There was a much more collaborative and enthusiastic sharing of ideas about phylogenesis between Ferenczi and Freud.[5] Nonetheless, Ferenczi was careful when disclosing his biological inclinations to Freud. In the period prior to 1919 he had professed his interest in biology but always in the context of reassurances that this did not constitute disloyalty to psychoanalysis: "I got lost in biological problems and can't find my way back to psychology! Fortunately, I know I am on the wrong track, and I hope—after leaving the mostly biological train—to land finally in the secure harbor of psychoanalysis" (Ferenczi and Freud 1996, 46). Again, a few months later:

It proved to be unavoidably necessary for me also to seek biological support for my hypotheses [about coitus]; for that reason I had to read embryological, zoological, and comparative physiological material. . . . I don't want to appear before you before I have emerged from biology and once again returned to the viewpoints of ψα. (To calm you, I will say right away that I have always conceived this detour only as a means to an end.) I hope that by the end of the month I will again be completely at home (with ψα) and will be able to listen to your new findings undisturbed by further distractions. (Ferenczi and Freud 1996, 60)

Four years later, in 1919, Ferenczi is still struggling to strike a balance between his speculations about biology and classical Freudianism. While acknowledging the fundamental Freudian tenet that conversion hysterias are the effect of unconscious wishes represented in the body, he wants the biological mechanism of hysteria to be more fully elucidated: "I believe that in spite of all our satisfaction with what has been achieved [in the metapsychology of hysteria], it is more to the purpose to indicate the lacunae in our knowledge of these matters. The 'mysterious leap from the mental to the bodily' (Freud), for instance, in the symptoms of conversion hysteria is still a problem" (Ferenczi 1919, 90–91).

When circumscribing his discussion of conversion to a particular class of bodily disruption, Ferenczi (1919) chooses not paralyses but afflictions of the gut. Globus hystericus (lump in the throat), for example, is one of the most common hysterical afflictions of the digestive tract. Ferenczi seems less interested in the interpretation of such a symptom (which he passes over in a perfunctory manner as "unconscious fellatio, cunnilingual, coprophagic phantasies" [92]) and more engaged with the material transformation that occurs in the throat: "The patients themselves speak of a lump stuck in their throats, and we have every reason to believe that the corresponding contractions of the circular and longitudinal musculature of the oesophagus produce not only the paraesthesia of a foreign body, but that a kind of foreign body, a lump, really is brought about" (92).[6] So, too, with anorexia nervosa, hysterical vomiting, and disruptions of the stomach and bowel. For Ferenczi, these symptoms demand an explanation of the conversion mechanism that is more conversant with gastric, intestinal, or esophageal substance. A special name is required for these events. He called them "materializations":

> From the analysis of a patient which is meeting with complete success, and from similar earlier and present observations, there arose a plan for a work on "hysterical materialization phenomena," especially in the gastrointestinal tract; an unbroken line from globus hystericus (fantasies of fellatio) over swallowing air (stomach), then to my case, which I mentioned, in which the patient was able at will to conjure up a penis in her vagina and a child in her intestines; finally; a nice case in a man of rectal "materialization" of a penis sticking into it. (All with the aid of tricks with the musculature of the intestine.) (Ferenczi and Freud 1996, 247)

Ferenczi quickly concedes that his use of the term *paraesthesia* in relation to hysterical symptoms is unwarranted. After all, these are not faulty perceptions or hallucinations on the patient's behalf; a material transformation really has been effected.[7] The increased capacity in hysteria for the fabrication of lumps in the throat or a child out of the contents of the stomach or a penis out of intestinal matter suggests that an aptitude for condensation, displacement, connotation, repetition, or identification cannot be contained to the ideational realm (dreams, parapraxes, and so forth); these capacities are also part of the nature of the body's organs, vessels, and nerve fibers. The action of the musculature of the intestines is not that of passive substrate awaiting the animating influence of the unconscious but, rather, that of an interested broker of psychosomatic events.

This vital contribution from the body's substrate guides Ferenczi to an important reformulation of the metapsychology of hysterical conversion. Materializations are not the effect of a leap from the mental to the somatic; rather, they are the product of a regression to a protopsychic state. Hysteria materializes the protopsychic (ontogenetic and phylogenetic) inclinations native to the body's substrata. By "ontogenetic tendencies" he means the desire to return to the womb (to fetal or embryonic conditions) in order to "bring about the reestablishment of the aquatic mode of life in the form of an existence within the moist and nourishing interior of the mother's body" (Ferenczi 1924, 54). Phylogenetic trends are the desire for all creatures to return to the water (the thalassal trend). As with the trauma of birth, terrestrial species have been traumatized by their expulsion from the water as the prehistoric seas receded. In this sense, ontogenetic and phylogenetic events are coeval: "What if the entire intrauterine existence of the higher mammals were only a replica of the type of existence which characterized that aboriginal piscine period, and birth itself nothing but a recapitulation on the part of the individual of the great catastrophe which at the time of the recession of the oceans forced so many animals, and certainly our own animal ancestors, to adapt themselves to a land existence, above all to renounce gill-breathing and provide themselves with organs for respiration of air?" (45). Ferenczi argues that these ontogenetic and phylogenetic inclinations (or motives) are latent in all substance but that they come to the fore most plainly in states of psychopathology. Such primal psychosomatic substrate is graphically illustrated in the 1932 clinical diary:

Inorganic and organic matter exist in a highly organized energy association, so solidly organized that it is not affected even by strong disruptive stimuli, that is, it no longer registers any impulse to change it. Substances are so self-assured in their strength and solidity that ordinary outside events pass them by without eliciting any intervention or interest. But just as very powerful external forces are capable of exploding even very firmly consolidated substances, and can also cause atoms to explode, whereupon the need or desire for equilibrium naturally arises again, so it appears that in human beings, given certain conditions, it can happen that the (organic, perhaps also the inorganic) substance recovers its psychic quality, not utilized since primordial times. In other words the capacity to be impelled by motives, that is, the psyche, continues to exist potentially in substances as well. Though under normal conditions it remains inactive, under certain abnormal conditions it can be resurrected. Man is an organism equipped with specific organs for the performance of essential psychic functions (nervous, intellectual activities). In moments of great need, when the psychic system proves to be incapable of an adequate response, or when these specific organs or functions (nervous and psychic) have been violently destroyed, then the primordial psychic powers are aroused, and it will be these forces that will seek to overcome disruption. In such moments, when the psychic system fails, the organism begins to think. (Ferenczi 1988, 5–6)

Ferenczi's analyses of extreme psychological duress substantiate a theory of the organic body that differs in important ways from Freud's hystericized body. In this extract from his clinical diary Ferenczi appears to distinguish between two kinds of psychosomatic organization. First, he recognizes the psyche-soma relation as it exists under normal conditions ("nervous intellectual activities" and organic matter "so solidly organized that it is not affected even by strong disruptive stimuli"). Pathologies in this organization are treatable according to the dictates of classical Freudian analysis: interpretation of ideational content as though anatomy did not exist. This body is neurotically inclined, symbolically guided, and analyzable.

Second, Ferenczi describes an organization of psyche and soma wherein primordial psychic powers emerge after normal psychic structures have been violently destroyed by trauma ("the organism begins to think"). Here, organic substance is intrinsically, primitively psychic ("impelled by motives"); psychological organization has regressed to a state where (as in the phylogenetic and ontogenetic past) it is not possible to distinguish matter from motive or deliberation. The psychotic disintegration consequent of severe trauma reveals to Ferenczi a bedrock of organic thought, and these extreme states provide him with the key to mechanisms of materialization in classical Freudian hysterias: "The hysterically reacting body could be described as semifluid, that is to say a substance whose previous rigidity and uniformity have been partially redissolved again into a psychic state, capable of adapting. Such 'semisubstances' would then have the extraordinarily or wonderfully pleasing quality of being both body and mind simultaneously, that is of expressing wishes, sensations of pleasure-unpleasure, or even complicated thoughts, through changes in their structure or function (the language of organs)" (Ferenczi 1988, 7).

At the end of the 1919 paper on hysterical conversion, Ferenczi notes that the conventional knowledge of human and animal physiology ("even the best and most exhaustive text-books" [103]) will not be adequate to the task of making the biology of hysterical materialization legible. These biological knowledges think of organs only in terms of their utility for the preservation of life. Instead he argues that biology must be approached "from the other side" (104)—that is, from the direction of psychoanalysis. If biological substrate was studied dynamically, the excessive concern with the utility (rationality) of organs that characterizes traditional biological knowledges could be supplanted with a more intricate account of their capacity for pleasure and destruction, for the expression of wishes, and for complicated thought. My ambition is not to take notions of "thought" and "motive" as we commonly understand them (narrowly cognitive) and simply apply them to the biological domain. Rather, I am hoping to denaturalize our habitual definitions of these terms by associating them with hysterical materialization. The thinking that an organism enacts when its cognitive, rational, symbolizing structures have been destroyed should provide an opportunity to reconsider the nature of thinking in the usual sense. Similarly, embedding motive or deliberation in biological substance is one way of broadening questions of causality beyond narrowly mechanistic definitions of organic influence.

In his 1924 book *Thalassa*, Ferenczi names this approach "bioanalysis" or "depth biology." *Thalassa* is perhaps Ferenczi's greatest accomplishment in terms of thinking biology differently, but it has not been widely read. I suspect this is because there are not yet enough conceptual tools available to help readers assess biological hypotheses as something other than wholly reductionist. Mechthild Zeul (1998), for example, is one of the rare Ferenczi commentators who does engage with the work in *Thalassa*, yet her reading conflates Ferenczi's bioanalysis with "biological concretism" (219). I am arguing that it is precisely a notion of biological concretism that Ferenczi's work (like Klein's) gorgeously animates; his bioanalysis is an attempt to bring depth and dynamism to conventional, two-dimensional ("flat") biological science so that it is no longer possible to automatically align a biological hypothesis with literal-minded reductionism.

What is at stake for me in this difference between a Freudian leap and a Ferenczian regression is not the question of whether hysteria is a movement forward (leaping) or backward (regressing), or whether it is a higher or a lower, complex or primitive psychological state. Rather, I am interested in how the notion of a leap invokes a gap of some sort between the mental and the somatic (a spatial divide between a psychic event and a bodily one that a conversion hysteria somehow, enigmatically, bridges) and, contrariwise, how Ferenczi's use of regression folds psychic events (from the present, the individual's past, and prehistory) into the heart of organic substrate. Without doubt, Ferenczi reinstates a division of psyche and soma elsewhere in his work. For example, even as he leads up to his reformulation of conversion as a regression, he keeps the question of a disjunction between ideation and (motor) substrate alive: "In the phenomenon of materialization . . . the unconscious wish, incapable of becoming conscious, does not content itself here with a sensory excitation of the psychic organ of perception, but leaps across to unconscious motility" (Ferenczi 1919, 97). This complex imbrication of loyalty to Freudianism with experimentation at the limits of psychoanalytic theory and technique structures most of Ferenczi's work (as it did Klein's). My goal is not to argue that Ferenczi avoids the conceptual traps to which Freud succumbs, but to map out how these major psychoanalytic thinkers struggle with an ontological puzzle that confounds us all.

By the 1930s the character of the organic had become largely inconsequential to traditional psychoanalytic theories of hysteria: analytic focus fell instead on the ideational contortions that hijack bodily function. The body had become the instrument of the unconscious rather than its symbiotic ally, and relatively little intellectual energy was given over to thinking about the nature of biological substance. More and more, the organic was envisaged according to the flat topology of conventional biological knowledge, even though Freud and others were exploring the strange vicissitudes of the hysterical body, and even though they were taking certain aspects of the body to be central to psychic structure. In this sense, classical psychoanalysis often distinguishes (even if only implicitly) between the bodily and the organic, the former being biological structures under the influence of the psyche, the latter being the biological residue liberated from or immune to such influence. The psychosomatic treatments that emerged out of psychoanalysis in the 1920s made this connotation in Freudian metapsychology explicit: the goal of psychosomatic practitioners such as Georg Groddeck, Felix Deutsch, or Franz Alexander was to rid the affected organ (e.g., liver, kidney, stomach) of its entanglement with the psyche. For example, Deutsch (who was a contemporary of Ferenczi and Freud's personal physician) renders organic questions immaterial to psychoanalytically inflected treatment: "The recurrence of organic symptoms during analysis is not surprising to the analyst, as he does not regard them as turns for the worse, knowing that the analytic treatment must finally render the organic expression of the conflict superfluous. What is really essential is to loosen psychosomatic ties, to purify the organs from their psychic cathexis, so to speak, and to assure their organic function unaffected by too strong libidinal forces" (Deutsch 1927/1964, 53). Ferenczi understood the organic and the analytic differently. Behind what he calls the façade of conventional biological description there is a biological unconscious. This biological unconscious motivates all organic activity; in certain (usually pathological) circumstances the phylogenetic and ontogenetic capacities that compose the biological unconscious come to "dominate the vital activities with their archaic impulses in the same way as the normal consciousness is inundated by psychological archaisms in the neuroses and psychoses" (Ferenczi 1924, 83). There is no way to purify organs of their psychic cathexis in the same way that there is no way to cleanse conscious cognitive processing of the influence of the unconscious.

For Ferenczi, the study of organic phenomena should connect at some point with a theory of the biological unconscious; while not all biological substrata are hystericized, a nascent kind of psychic action (motivation, deliberation) is nonetheless native to biological substance. In the same way that Freud used hysteria to reveal the neurotic/phantastic nature of the normally functioning psyche, Ferenczi uses an analysis of materialization to reveal the plastic nature of all organic substrate ("I do not believe that we are dealing here with processes that hold good for hysteria only and are otherwise meaningless or generally absent" [Ferenczi 1919, 103]). In so doing, he generates a schema for feminists wanting to think about biological substrate as another scene, rather than as bedrock.

There has been a tendency (largely unrecognized) in feminist theory to act out this troublesome distinction between the bodily and the organic. It is not simply that there has been a preference for encountering embodiment via social, representational, or symbolic analysis at the expense of biological data (see the introduction and chapter 1). This, after all, could simply be an effect of disciplinary affiliation, correctable by an increase in the number of feminists writing with an interest in the biological. More problematically, many feminist theorists seem to gesture toward a flat organic realm elsewhere as a way of securing a more valuable or dynamic account of politics closer to home. The organic—conceptually dull and politically dangerous—lurks at the periphery of these texts (e.g., Rubin), underwriting the claims about embodiment that are made.

Ferenczi provides one way through this impasse. Under Ferenczi, biology is strange matter, proficient at the kinds of action (regressions, perversions, strangulations, condensations, displacements) usually attributed only to nonbiological systems. This biology is not the flat (sovereign, authoritative, juridical) substrate seen in many feminist or neuro-humanities arguments; it is much less tractable to conventional empiricism and politics. Clearly, hysteria comprehends more about the body than just what is given by perceptual and tactile data; hysteria also enacts some knowledge of the biological unconscious—the ontogenetic and phylogenetic impulses that motivate the body's substrata. Conversion is an immediate and intimate psychosomatic event. It is not an ideational conflict transported into the bodily realm; it is not the body expressing, representing, or symbolizing a psychic conflict that originates elsewhere. To return to an early formulation of hysteria in the opening paragraphs of this chapter: conversion hysteria does not point to what is *beyond* the organic body. On the contrary, conversion hysteria is a kind of biology writing (Kirby 1997), and as such it shows the dynamic character of all organic matter.

Antiperistalsis

Marya Hornbacher was nine years old when she first began to induce vomiting. By the time she was fifteen years old, her habit had become complicated by anorexic periods and drug abuse. At this time, she reports, "I was hardly eating anything at all. Rice, bits of fish. I perfected the art of the silent puke: no hack, no gag, just bend over and mentally will the food back up" (Hornbacher 1998, 97). Hornbacher's memoir of her Eating Disorder Not Otherwise Specified illustrates the chaotic nature of contemporary materializations in the gut. They are often comorbid with each other (bulimia and/or anorexia and/or binge eating), and they are often complicated by a variety of other symptoms of distress (anxiety, depression, disordered personality, substance abuse). They are disproportionately diagnosed in women, and in the current popular psychology parlance they are taken to be acting out "issues of control." The mental determination of people with eating disorders is commented on frequently, and Hornbacher folds this characteristic into her description of her routine vomiting ("mentally will the food back up"). It is by no means clear, however, that the bulimic accomplishment of habitual vomiting can be attributed solely to ideational capacities (mental will, remote from and in control of the gut). What might be the Ferenczian character of this "willful" practice? What feminist readings of mind-body can be fashioned using data from the pharmaceutical treatment of eating disorders?

The capacity to will food back up is commonly developed in bulimia. The DSM-5 notes this skill as a matter of course: "Individuals generally become adept at inducing vomiting and are eventually able to vomit at will" (APA 2013, 346).[8] Gerald Russell (who in 1979 first laid out the diagnostic criteria for bulimia nervosa) notes two variations of this skill: patient 1, in whom vomiting had become effortless ("I just had to think about it. I don't have to put my fingers down my throat. I press my stomach and I am sick" [Russell 1990, 19]); and patient 2, who at first induces vomiting by putting her fingers down her throat, but "in later years she simply drank some fluid and leaned over the toilet so that 'it all came up in one go'" (20). Paul Robinson and Letizia Grossi in a 1986 letter to the *Lancet* observe that the gag reflex itself may be attenuated in bulimics:

> In studies of gastric emptying and oesophageal pressure in patients with bulimia nervosa, we have asked controls and patients to accept a nasogastric or an orogastric tube. In four normal controls (including ourselves) this procedure was unpleasant; it took 10–15 minutes and was accompanied by retching and lacrimation. One control passed the tube without difficulty. We approached the patients cautiously, warning them to expect these symptoms. To our surprise, six of the seven bulimic patients had no difficulty swallowing the tubes, doing so in seconds with no sign of distress. In the patients a gag reflex could not be elicited by stimulation of the fauces [back of throat]. One patient who did have a gag reflex on swallowing the tube was a bulimic patient who had not induced vomiting for three years. (221)

To say that these phenomena are attributable to the reconditioning of a reflex action is to beg the Ferenczian question. Is the gag reflex a simple mechanical action distinct from psychic or deliberative impetus? Does its disconnection from higher cortical centers (and so from conscious cognitive processing) render it a nonpsychological event? What is conditioning anyway? It seems to me that the gag reflex, this seemingly rudimentary biological action, is a very useful place from which to start thinking about the organic character of disordered eating. Freed from an immediate concern with the ideational and cultural systems that help enact such a symptom, we are able to observe the vicissitudes of organic thought. Russell's second patient, who uses ingestion (drinking) to provoke vomiting, is not simply perverting the course of normal peristalsis; nor is she simply and in a mechanical way reconditioning a hardwired (flat) nervous event. Rather, the soft tissue at the back of her throat (as with Robinson and Grossi's bulimic patients, and with Hornbacher) has become alive to a number of different ontogenetic and phylogenetic possibilities (i.e., to what Ferenczi calls the biological unconscious).

Here standard anatomical texts help orient us to the primal nature of the throat's substrate. *Gray's Anatomy* describes the fauces as "the aperture by which the mouth communicates with the pharynx" (Gray 1918, 1137). The pharynx connects at the upper end with the mouth, nasal passages, and ears and at its lower end with the esophagus. Moreover, the pharyngeal area is "the embryological source of several important structures in vertebrates. For example, the breathing apparatus (gill pouches of fish and lungs of land animals) arises in this area. . . . In humans, the pharynx is particularly important as an instrument of speech" (Lagassé 2000, n.p.). The back of the throat is a local switch point between different organic capacities (ingestion, breathing, vocalizing, hearing, smelling) and different ontogenetic and phylogenetic impulses. Much more than the front of the mouth or even a little lower down into the esophagus itself, the fauces is a site where the communication between organs may readily become manifest.

Ferenczi has sketched out such interorgan communication at another overdetermined biological site. In *Thalassa*, he hypothesizes that the anal and urinary organs are intimately entangled in terms of function: "The organs participating in urethral functioning are crucially influenced from the anal sphere, the organs of anal functioning from the urethral, so that the bladder acquires a degree of retentiveness from the rectum, the rectum a degree of liberality from the bladder— or scientifically stated, by means of an amphimixis [mingling] of the two erotisms in which the urethral erotism receives anal admixtures and the anal erotism urethral" (Ferenczi 1924, 12). The rectum communicates its retentiveness to the bladder; the bladder communicates its liberality to the rectum. Without such interorgan exchange, the bowel would become hopelessly constipated and the urinary tract incontinent. Amphimixis is not a secondary perversion of flat biological substrate; it is the very means by which these organs are able to function naturally at all. This anal-urethral admixture spills over into the copulative act, such that the genitals (for Ferenczi, at this moment, the penis) acquire their natural function (ejaculation) through amphimixic relations to the bladder and bowel: "The genital would then no longer be the unique and incomparable magic wand which conjures eroticisms from all the organs of the body; on the contrary, genital amphimixis would merely be one particular instance out of the many in which such fusion of erotisms takes place" (Ferenczi 1924, 12).

So, too, at the other end of the digestive tract, various organs of ingestion, expulsion, sensation, and expression are borrowing from one another, but this time under the pressure of pathology. In Russell's patients, the gagging capacities of the fauces have borrowed from the pharynx and become more like swallowing, and ingestion has become a technique for expulsion rather than digestion. Here the quality of organic amphimixis is more acute and chaotic than in Ferenczi's case of the urethra, rectum, and genitals. A more detailed examination of any one case of bulimia would no doubt find a trade in unconscious inclinations among the digestive organs and their neighbors and between ontogenetic and phylogenetic registers: "Once attention is directed to [the biological unconscious] it will certainly become possible to recognize more definitely in certain anomalies of nutrition—in its pathology, for example—the activation of regressive tendencies which under ordinary circumstances remain hidden. In such fashion one would perceive behind the symptom of *vomiting* not only its manifest immediate etiology but also tendencies towards regression to an embryonic and phylogenetic primevality in which peristalsis and antiperistalsis were mediated by the same digestive tube" (Ferenczi 1924, 86).

The longer bulimia continues, the more manifest and routine this organic thought becomes. As Hornbacher's memoir demonstrates, episodes of bingeing and vomiting in chronic bulimia are often no longer directly tied to meaningful, analyzable events in the patient's internal or external world. Bingeing and vomiting can become compulsive, and it is for this reason that some commentators want to figure bulimia as an addictive disorder. By the time the bingeing and vomiting of bulimia have become functionally autonomous, bulimia is extremely difficult to treat: the organism itself is beginning to think. Distress, anger, need, depression, comfort, and attachment have become primarily organic, and their capacity to respond to cognitively or ideationally oriented treatments is greatly reduced.

My argument is that the bulimic capacities of the throat should draw our attention not just to behavioral intent (will) or cultural transformation or disorder in higher cortical centers or mechanisms of unconscious representation but also to the Ferenczian action of the digestive organs. The vicissitudes of ingestion and vomiting are complex thinking enacted organically: bingeing and purging are the substrata themselves attempting to question, solve, control, calculate, protect, and destroy.

Gut Analysis

One of the ways in which chronic bulimia can be treated is via the administration of antidepressants. Since the 1970s it has been clear that a variety of antidepressant medications (tricyclics, MAOIs, and SSRIs) can significantly reduce bingeing and purging in bulimic patients. Over reasonably short periods of time (six to sixteen weeks), double-blind, placebo-controlled clinical trials have demonstrated that binge episodes significantly decrease (although they are usually not entirely eliminated) when subjects are on a course of antidepressant medication: "The mean reduction in binge-eating frequency across studies ranged from 22% to 91% with reductions most often reported in the 50% to 70% range. Abstinence rates at the end of treatment have ranged from 0% to 68%, with a mean of 24%" (Mitchell et al. 2001, 298).[9] Harrison Pope and James Hudson's (1986) review of drug therapy for bulimia assessed the efficacy of a wide variety of thymoleptic (mood-stabilizing) medications: for example, lithium carbonate, phenytoin (an anticonvulsant), and even methamphetamine. Early research on most of these drugs was inconclusive; it was the antidepressants that emerged as the most reliable form of treatment for bulimia.

Nonetheless (and this is the point I would like to exploit), it is unclear why antidepressants are so effective in the treatment of bulimia; there is no agreement in the literature as to how the relationship between mood and bingeing should be understood. Two main schools of thought about the pharmaceutical connection between bulimia and depression have emerged. First, some researchers (e.g., Pope and Hudson 1986) use the data to argue that bulimia is, in fact, a mood disorder. That is, bulimia is a variant of depression, and if depression is treated successfully, the bulimic symptoms will likewise fade away. The fact that there is a high comorbidity of depression in bulimia supports this thesis. However, the character of depression in bulimic patients seems to be of a different quality from that of typical major depression: the diurnal variations in mood, diminished libido, irritability, and lack of concentration that constitute many depressions are often absent in bulimic patients, even though they may have depressed affect and be suicidal (Russell 1979). Moreover, it is difficult to diagnose depression in bulimia, as bingeing and purging disrupt or mask the usual biological markers of depression, such as weight gain or loss, decreased energy, and disturbance in gastrointestinal function. A second explanation for the effectiveness of antidepressants in reducing bulimic symptoms is that these pharmaceuticals have a direct effect on appetite, independent of their antidepressant effects. Data showing that fluoxetine hydrochloride (Prozac) alleviates bingeing and purging, irrespective of whether or not the patients are depressed, support the hypothesis that these pharmaceuticals are acting directly on satiety mechanisms in the brain (e.g., Goldstein et al. 1995; Leibowitz 1990). For these researchers, bulimia is first and foremost a pathology of eating, to which depression may or may not become attached.

Despite over thirty years of clinical research on bulimia and antidepressants, there is no clear biomedical etiology for bulimia: Does depression trigger bulimia? Or do pathologies in the brain's satiety mechanisms instigate disregulated eating, provoking severe dieting and eventually bingeing and purging—all of which are independent of mood? There are a number of demarcations that these etiological discussions in the literature seem to force on the reader: depression *then* bingeing; satiety *or* mood; brain *not* gut. It has been my argument, via Ferenczi, that these Boolean demarcations among organs and between psyche and soma are intelligible only within a conventional (flat) biological economy. It seems to me that the lack of a clear path from one cause to one effect, from one organ to another, or from the psychological realm to the biological and back again, indicate not a lack of conclusive data but the workings of the biological unconscious made manifest. Perhaps the lability of eating and mood—their tendency to align and dissociate under the influence of certain medications—speaks to an ontological organization that is at odds with organic rationality. As is so often the case in contemporary biomedical literatures, there is an overriding concern with clearly demarcating causal primacy (what causes what?)—as if determination is a singular, delimited event. The limitations of such an approach to biological explanation were evident to Ferenczi:

> This seeing things only in the flat, so to speak, had the result that in the natural sciences one was satisfied, in general, with a conception of vital phenomena limited to a single interpretation of the data. Even psychoanalysis was not so long since committed to the view that it was the prerogative of the psychic sphere alone that its elements, indeed one and the same element, could be inserted *simultaneously* into several genetically different causal series. Analysis expressed this fact by the concept of *overdetermination* of every psychic act, as the direct consequence of the polydimensional character of things psychic. Just as at least three coordinates are necessary in order to define a point in space, so in the same way neither a psychic datum nor, as we indeed believe, a datum in the field of physical science is sufficiently determined by its insertion in either a linear *chain* of events or in a two-dimensional *nexus* thereof, unless its relationships to a *third dimension* [the biological unconscious] are also established. (Ferenczi 1924, 84)

The bingeing and purging of bulimia, and their alleviation by the administration of antidepressants, are not explainable until, at the very least, a connection has been made to organic thought and to the amphimixic inclinations of the substrata involved—that is, until a more dynamic schema for digestion, respiration, antiperistalsis, neurotransmission, and mood has been established.

Rather than seeing a lack of pattern in the clinical data, I see support for Ferenczi's thesis of a protopsychic substrate that is capable of differentiated, phantastic action (a lump in the throat, a child in the stomach, a penis in the rectum). The gut is sometimes angry, sometimes depressed, sometimes acutely self-destructive; under the stress of severe dieting, these inclinations come to dominate the gut's responsivity to the world. At these moments any radical distinction between stomach and mood, between vomiting and rage, is artificial. Here a clear indication of what is meant by *radical* (pertaining to the root: foundational, essential, originary, primary) is important. I am not arguing that organs are indistinguishable from one another, or that psyche and soma are the same thing. Rather, I am claiming that there is no originary demarcation between these entities; they are always already coevolved and coentangled. For this reason the routine critical response that bulimic etiology can be attributed to an interaction (mind + body) is inadequate for the argument I wish to make here. The logic of interaction, addition, or supplementarity presumes that the entities at stake are already detached and can be brought into a relation for the purposes of specific pathologies (Barad 2007; Kirby 2011; Oyama 2000). I am arguing that antidepressants alleviate bulimia because there is no radical (originary) distinction between biology and mood. Mood is not added onto the gut secondarily, disrupting its proper function; rather, temper, like digestion, is one of the events to which enteric substrata are naturally (originally) inclined. Manfred Fichter and Karl Pirke (1990) allude to such a psychosomatic structure when they conclude a discussion of endocrine dysfunction in bulimia nervosa by suggesting that, in addition to thinking of disruptions to eating as symptoms of depression, it may also be useful to think of depression as a kind of nutritional disorder.

The clinical data on bulimia and antidepressants indicate extensive traffic among the body's organs and between the gut and mood in ways that are not delimitable to a flat logic of biological matter. The notions of amphimixis, of a biological unconscious, of materializations and organic thought surely breach the boundaries of respectable biological theory, but they do so without gesturing to antibiological ground. These Ferenczian provocations, in conjunction with more recent empirical studies, suggest that depression is a biologically distributed phenomenon: it cannot be explained only with reference to the central nervous system or only in relation to cognitive distortion. For example, fluoxetine hydrochloride does not just act centrally (on, say, the serotonin pathways in the hypothalamus, which are thought to administer eating) and cognitively (to reorient infelicitous thinking); it also acts peripherally, in the gut. Most of the body's serotonin (about 95 percent) is to be found in the complex neural networks that innervate the gut (Wilson 2004). While this is not usually discussed in the psychiatric literature, the gut itself (the stomach and attendant viscera, and their specific modes of organic deliberation) are being soothed (or in some cases agitated) by serotonergic treatments. That is, antidepressants do not have effects on mood simply by influencing the brain; they also directly enliven the viscera—in the case of bulimia, calming distress that is now more enteric than cerebral in character. The responsivity of bulimia to anti-depressants is one key piece of data that illuminates psychic action in the gut—its phantastic capacity to digest and ruminate.

Conclusion

Chapters 4, 5, and 6 expand on the details of serotonergic action in the body (pharmacokinetics and transference; the work of placebo; pharmaceuticals and suicidality), and under the sway of the Ferenczian logic outlined here, they use pharmaceutical data about depression to think about the mindedness of the viscera and the worldliness of biochemical action. But first, the next chapter intensifies the critique of feminist theory that has begun here. In addition to the difficulties that feminist theory has had grappling with biological data, there is a widespread aversion to thinking about aggression as a key part of a depressive scene and a vital part of political activity. What are the hostilities, aggressions, and animosities that make feminist politics possible? Might we begin to think of aggression not simply as the harmful action of others, but as the necessary condition for every feminist engagement?

CHAPTER 3

BITTER

MELANCHOLY

Traditionally, depression has been conceived of as the response to—or expression of—loss. . . . The hostility that should or could be directed outward in response to loss is turned inwards towards the self.
—Phyllis Chesler, *Women and Madness*

One of the most common explanations of depression is that it is anger turned inward. The aggression, hostility, indignation, or rage that ought to be expressed toward others is stifled and turned back onto the self. This has been a particularly popular explanation for depression in women. As Phyllis Chesler notes in her early, influential book *Women and Madness*, " 'Depression' rather than 'aggression' is the female response to disappointment or loss" (1972, 42). Because women are more likely to suffer emotional and economic losses than men, and because these losses receive insufficient social recognition or restitution, the anger of women sometimes morphs into a toxic internal state of self-reproach, hopelessness, and guilt. It is for this reason, it would be argued, that women are two to three times more likely to be diagnosed with a depressive illness than men.[1]

This explanation about the inward-turning character of depression is usually attributed to Freud. In his canonical essay "Mourning and Melancholia" Freud argues that a lost object is never entirely abandoned libidinally. In cases of melancholia the libido that was previously attached to the object is withdrawn into the ego, and this instigates an internal scene where the ego identifies with the abandoned object: the shadow of the object falls upon the ego (Freud 1917a). From this point on, the melancholic's internal world is colored in new ways by the lost/ loved object. In particular, because the original attachment to the object was likely ambivalent (the object was both loved and hated), melancholia amplifies not only grief but also the hatred already felt toward the object. This ambivalence isn't the effect of the loss; rather, it is the milieu within which melancholic losses happen. That is, the melancholic hates the object she loves, and she does so prior to the object's loss. It is this complex set of emotional and unconscious events that emerges in later commentaries as the pared-down hypothesis of anger-turned-inward. Ann Kring, Sheri Johnson, Gerald Davison, and John Neale's (2010) psychology textbook, for example, summarizes Freud's account in this way: "Freud asserted that the mourner unconsciously resents being deserted and feels anger toward the loved one for the loss. . . . The mourner's anger towards the lost one becomes directed inwards, developing into ongoing self-blame and depression. In this view, depression can be described as anger turned against oneself" (231).[2]

There are a number of simplifications and confusions in these post-Freudian accounts of depression: there is a tendency to treat sadism as anger, when they are not the same thing psychologically; the primacy of hatred in psychic life is often underestimated, or overlooked entirely; sadism ends up sequestered inside the melancholic (as private self-hatred), so the ways in which hostility is perpetrated by the self against others remain underexplored.[3] In this chapter I will consider the relation between depression and hostility in more depth, and discuss the importance, for feminist theory, of thinking about depression as a kind of aggression *directed outward*. I will argue that the popular anger-turned-inward hypothesis makes depression look much less sadistic toward others and much more like self-enclosed self-hatred than Freud (and his collaborator on depression, Karl Abraham) intended.

Many feminist and critical accounts of loss have relied on Freud's theory of melancholia and the trope of an inward turn (e.g., Butler 1997), yet they have shied away from the vicissitudes of destructive melancholic feelings and actions that target the world (Balsam 2007). One key problem is the status of ambivalence (love and hate) in these analyses. David Eng and David Kazanjian (2003), for example, argue that ambivalence is a consequence of loss: "Melancholia results from the inability to resolve the grief and ambivalence precipitated by the loss of the loved object, place, or ideal" (3). For Eng and Kazanjian, ambivalence is triggered by loss rather than being (as Freud argues) the precondition that turns any loss melancholic. For them, hostility is provoked, not anticipatory. By this interpretation, hatred and sadism are (possibly reasonable) responses to disappointment, bereavement, or injustice, rather than (irrational and destructive) dispositions that we bring to bear on the world from the start. This placement of ambivalent feeling may seem conceptually minor, but it shapes a political orientation within which the primacy of our own hostility is disavowed, and it establishes an expectation that sadism toward the objects, places, and ideals to which we are attached can be avoided or eradicated. I will be interested in the ways that feminist politics can be rearranged if we begin with the assumption that loss is not simply something that happens to the melancholic but that losses are also one of the things that melancholy necessarily, spitefully enacts (wrecking bonds, attacking objects). If aggression and destruction are an enduring part of any depressive scene, then how does feminist theory orient itself to that bitterness?

This bitterness is already contained in the history and etymology of melancholy: black *bile*. This draws my interest in depressive hostility to the internal, digestive organs (rather than, say, gender differences in depression).[4] As will be clear by now, the biology that I am pursuing is not a flat, sovereign substrate. Rather, I am interested in biological events that show evidence of a biological unconscious—a biology that could be in a forlorn and destructive relation to itself and the world. I classify these concerns as feminist because they push conventional understandings of mind-body to the limit: Can the viscera be minded? Can biological substrata be melancholic, aggressive, bitter? In particular, it seems to me that feminist theory has more room to engage the current psychopharmaceutical milieu (which will be the concern of the last three chapters of this book) when it understands that biology has as a part of its nature a constitutive bitter hostility.

After Eve Kosofsky Sedgwick's (1997) influential essay on paranoid reading, there has been a lot of enthusiasm for moving theory into a reparative (nonbitter) mode. There has been a tendency (in both Sedgwick's original essay and her subsequent readers) to put paranoid reading on the side of aggression and reparative reading on the side of consolation, affirmation, amelioration, or the reconstruction of a sustainable life (Berlant and Edelman 2013; Hanson 2011; Love 2010;

Wiegman 2014). This tends to bifurcate the political and conceptual field too agreeably. The problem, as I see it, is not just that any particular reading (starting with Sedgwick's) can be both paranoid and reparative, by turn. The real difficulty is the tendency to figure reparation as nonaggressive.[5] It is my presumption in this chapter that calls for a reparative approach to reading cannot be calls to do away with bad objects or hostility (see the conclusion). Reparation, in the Kleinian sense, is an attempt to repair the damage done by our attacks on objects. Reparation may involve concrete acts in the world, but more likely it is enacted in phantasy. It is not clear that such reparative acts are ever successful; they are likely to be partial or compulsive or temporary, and certain kinds of repair damage objects further (Spillius et al. 2011). As Esther Sánchez-Pardo (2003) notes: "Reparation and integration are only partial outcomes, never fully achieved, which bring about various ways of living with damage" (133). Crucially, reparation is directed toward damaged objects; it is not an attempt to abolish sadism itself. In fact, I would argue that reparative gestures require, first of all, the recognition that sadistic attacks are inevitable, that they can originate from me, and that while their viciousness can be down-regulated, such attacks cannot be eradicated: the best we can hope for is that "a relative balance between love and hate is attained" (M. Klein 1940/1975, 351). Damage to others is an inevitable effect of being in relations with others. If we are compelled to repair the damage we do, this happens without the possibility of final redemption or restoration: "The unification of external and internal, loved and hated, real and imaginary objects is carried out in such a way that each step in the unification leads again to a renewed splitting of the imagos" (M. Klein 1935/1975, 288).[6]

The nature of attacking, sadistic impulses, and the difficulties of how to live (and politick) with them, have not yet received sustained attention in feminist theories of depression (or reparation). It is not my goal here to imagine a specifically feminist hostility, because the modifying effect of "feminist" on "hostility" will just about always come to mean aggression that, in the end, bends to the good. Rather, I want to instigate some curiosity about a depressive hostility that has no ambition, and no other trajectory, except the destruction of the objects that the depressive loves, and to argue that acknowledgment (and perhaps some appreciation) of sadistic destructiveness is a necessary part of any feminist account of melancholic scenes.

Insatiable Sadism

Everything [melancholics] say about themselves is at bottom said about someone else.
—Sigmund Freud, "Mourning and Melancholia"

Let me begin by briefly outlining the salient parts of Freud's account
of melancholia: What place does sadism have in the formation of a depressive stance? Importantly, how internally contained is the hostility
that Freud says is a key part of a melancholic response? Toward the end
of this section, I will start bending these comments toward a consideration of the gut.

The first thing to note, Freud argues, is that melancholia (depression) is a different kind of response to loss than mourning (grief).[7]
Mourning and melancholia are similar in that they are both painful,
they both involve a loss of interest in the world, and the capacity to love
or be attached is greatly diminished in each case. What distinguishes
melancholia from mourning, and what marks melancholia as a pathological reaction, is a significant lowering of self-regard (which is not
present in states of grief). The melancholic hates herself: "In mourning it is the world that has become poor and empty; in melancholia it
is the ego itself. The patient represents his ego to us as worthless, incapable of any achievement and morally despicable; he reproaches himself, vilifies himself and expects to be cast out and punished" (Freud
1917a, 246). It is this initial part of Freud's description of melancholia
that seems to have been taken up as the hypothesis that depression is
anger turned inward. However, Freud's explanation of *why* the melancholic hates herself brings to light a much more complicated (and
compelling) set of circumstances than just a turn against oneself. His
key insight is that the self-reproaches of the melancholic are not accurate accounts of the moral failings of the patient. On close inspection
it seems that the self-reproaches are, in fact, rebukes against someone
else: "If one listens patiently to a melancholic's many and various self-
accusations, one cannot in the end avoid the impression that often the
most violent of them are hardly at all applicable to the patient himself,
but that with insignificant modifications they do fit someone else,
someone whom the patient loves or has loved or should love" (248). In
the first instance, then, we need to be inquisitive about the performa-

tive nature of depressive utterances. If self-reproaches don't simply describe a state of affairs (I am worthless), then what work are they doing? In the end, who is the target of melancholic self-vilification?

While Freud notes that accusations against others have been shifted onto the self, he is also clear that the hostility of this maneuver is not contained within the patient: "[Melancholics] make the greatest nuisance of themselves, and always seem as though they felt slighted and had been treated with great injustice. All this is possible only because the reactions expressed in their behaviour still proceed from a mental constellation of revolt" (248). The aggression of melancholia is not held privately; it is also being directed against others. The melancholic is in revolt. But likely she is not remonstrating in a heroic kind of way (where wrongs can be set to right, or where structures of oppression can be destabilized or overthrown). This is not resistance in the sense that it was envisaged by some early feminist commentators on hysteria ("Dora seems to me to be the one who resists the system. . . . And this girl—like all hysterics, deprived of the possibility of saying directly what she perceived . . . still had the strength to make it known" [Cixous and Clément 1985, 285]). Rather, the melancholic is in revolt, quite precisely and destructively, against the object she loves: "The loss of a love-object is an excellent opportunity for the ambivalence in love-relationships to make itself effective and come in to the open" (Freud 1917a, 250–251). The self-hatred of depression is an unconscious hatred of someone else, and the nuanced development of a sadistic stance toward them.

There is a subtle, but crucial, difference between the hypothesis of anger-turned-inward and the hateful ambivalence of melancholia. In the former case, anger fails to find expression in the world; it has been obstructed, turned around, and then it corrodes the self. This sequence of events seems to imply that if anger were expressed (brought out into public as militancy, for example), the corrosive effects on the self would be brought to an end. This perhaps more accurately represents an earlier phase of Freud's work: his studies on hysteria. With Joseph Breuer, Freud argued that a method of abreaction (emotional catharsis) would resolve symptoms of hysteria: "We found, to our great surprise at first, that *each individual hysterical symptom immediately and permanently disappeared when we had succeeded in bringing clearly to light the memory of the event by which it was provoked and in arousing its accompanying affect, and when* the patient had described that event in the greatest possible detail and had put the *affect into words. . . . [This method] brings to an end the operative force of the idea which was not abreacted in the first instance, by allowing its strangulated affect to find a way out through speech*" (Freud 1893b, 255). However, with the development of a theory of the unconscious after 1900 Freud no longer argued that symptoms can be abreacted away, and his account of the dynamics of distress became more complex and the politics of depression correspondingly more entangled (love with hate, aggression with pain, inside with out). In the case of melancholia, hostility hasn't been strangulated; on the contrary, hostility has found its mark: "The patients usually still succeed, by the circuitous path of self-punishment, in taking revenge on the original object and in tormenting their loved one through their illness, having resorted to it in order to avoid the need to express their hostility to him openly" (Freud 1917a, 251). The turning of aggression onto the self doesn't bring sadism toward the other to an end.[8] If I turn my hostility onto myself, I do not therefore relieve you of its effects. Rather, I transform those effects into different forms—forms that are harder to identify as outward aggression, but no less sadistic for their indirect character. Eng (2000) makes the inward turn of aggression too complete (melancholia "transforms *all possible* reproaches against the loved object into reproaches against the self" [italics added; 1276]), and this makes melancholia a closed system ("a turning away from the external world of the social to the internal world of the psyche" [1276]). Here I am wanting to emphasize that aggression turned outward is a crucial part of the traffic between ego and world, and that we need to keep this mode of relationality alive in theories of depression.

It is important to remember, at this juncture, that melancholia can be instigated not only by severe losses—like the death of a loved one or the destruction of one's community—but also by less immediately brutal situations: being slighted, neglected, disappointed. It is these latter circumstances that make melancholia quotidian—the kind of everyday dysphoria that has been exploited by pharmaceutical companies since Prozac. In this way, the hostility of depression may be dispersed through the everyday world as small shards of unconscious aggression (dysthymia) rather than an all-out assault on the other. In either case, this is no longer a picture of righteous (and conscious) anger gone astray. Rather, melancholic responses to loss are often malicious and usually impossible to like. This is a scene of intense bitterness—simultaneously painful and enjoyable. The depressive is both miserably desperate for attachment and elated by the destruction of the objects for which she yearns: "The self-torment in melancholia, which is without doubt enjoyable, signifies . . . a satisfaction of trends of sadism and hate which relate to an object" (Freud 1917a, 251). It is this unpalatable aspect of melancholia, I will argue in later sections, that is missing in feminist accounts of depression and melancholia.

In the meantime, I want to draw this argument about aggression closer to biology. These melancholic trends of sadism and hate, which find their way to others via self-torment, often materialize in the gut. Freud had been aware of a connection between gut and depression decades earlier. In a draft on melancholia in 1892, he notes the close relationship between disordered mood and disordered eating: "The nutritional neurosis parallel to melancholia is anorexia. The famous anorexia nervosa of young girls seems to me (on careful observation) to be a melancholia" (Freud 1892, 200). At this point Freud sees an affinity between anorexia and melancholia, but he has not yet come to a more finessed account of the constitutional aggression of these disorders. Melancholia is painful, but not yet hostile. It was his colleague Karl Abraham who first brought psychoanalytic attention to the importance of sadism in depressive symptomology: "In these persons [melancholics] an insatiable sadism directed to all persons and all things has been repressed" (Abraham 1911, 146). Abraham also saw an intimacy between melancholia and alimentary events that was infused with hostility:

> Many neurotic persons react in an anal way to every loss, whether it is the death of a person or the loss of a material object. They will react with constipation or diarrhoea according [to how] the loss is viewed by their unconscious mind. . . . News of the death of a near relative will often set up in a person a violent pressure in his bowels as if the whole of his intestines were being expelled, or as if something were being torn away inside him and was going to come out through his anus. . . . We must regard the reaction as an archaic form of mourning. (Abraham 1924, 426)

Abraham called this violent pressure in the bowels "organ speech." Freud had first used the phrase *organ speech* to describe hypochondriacal attacks. In his essay on the unconscious, he refers to a patient of Victor Tausk who was brought to the clinic after a lover's quarrel, complaining that her eyes were twisted. She claimed that her lover was a hypocrite, an eye-twister [Augenverdreher/deceiver]: "He had twisted her eyes; now she had twisted eyes; they were not her eyes any more; now she saw the world with different eyes. . . . I agree with Tausk in stressing in this example the point that the patient's relation to a bodily organ (the eye) has arrogated to itself the representation of the whole content [of her thoughts]" (Freud 1915, 198). For Freud, the body's organs are a proxy for, or representation of (Vertretung), the content of her thought. Despite his often nuanced accounts of the body of neurotics, this is a fairly conventional model of mind-body relations. There is a clear distinction, in this particular case, between the nature of organs and the actions of mind: organs can be vehicles for thought, but they are not themselves capable of deliberation. Her eyes take on the burden of representing deception or twisting, but for Freud and Tausk the mechanism of her distress is to be found in ideational or affective distortion in the unconscious. Mind acts, organs follow. Here I am wanting to expand the meaning of organ speech to make the entanglement of psyche and soma more explicit and to bring the psychically animated nature of biological substrata to the fore.

Could we think of organ speech as a kind of bodily utterance, in the sense meant by J. L. Austin (1962)? Perhaps organ speech is a biological performative—it enacts the events it appears only to be symbolizing. That is, the bowels are doing the archaic work of mourning. Constipation and diarrhea are not actions auxiliary to a state of grief, nor does it make any sense to say that these are constative acts (true or false). Rather, the savage responses of the bowel (to seize up or to relinquish destructively) are modes of grief, enterically performed. Specifically, the action of the bowels (a violent pressure as if the whole intestines were being expelled) is an aggressive kind of biological mourning—turned on the self yet also directed to the lost/loved, phantastic/physiological, internal/external, incorporated/masticated object. The melancholic's hatred of the object, now inflamed phantastically and physiologically, takes the form of violent enteric rejection: the object is being torn away and anally expelled. As Ferenczi might suggest, in the wake of damage both large and small, the digestive organs begin to deliberate, lament, and destroy.

In a similar vein, Abraham (1924) reports the case of a young melancholic man who "had a compulsion to contract his *sphincter ani*" (443). Abraham takes the symptom to be overdetermined, symbolically: "[It] stood for a retention . . . of the object which he was once more in danger of losing . . . [and also] his passive homosexual attitude towards his father" (443). My immediate concern here is not the possible reductiveness of the interpretations (anal retentiveness; the negative Oedipus complex), but rather the restriction of overdetermination to the ideational field. In concert with psychoanalytic orthodoxy (then and now), Abraham sees the action of the sphincter ani as a manifestation of conundrums that are being enacted elsewhere, in the psychic realm. For him, it is the ideational contortions of the unconscious that are overdetermined; the contracting sphincter muscles are merely the storehouse for those phantastic events. In these conventional psychosomatic interpretations, the actions of the sphincter are related to minded states only secondarily, under the sway of neurotic ideation. I am in search of ways to think about the sphincter muscles (and the walls of the bowel) as sources of mind, not merely instruments for the expression of feeling. Isn't it possible (if we think away from the flat topography of conventional biology and toward biological phantasy) that the compulsion to contract the sphincter ani is a psycho-enteric act? What if object loss is being felt and managed right here, by the sphincter muscle itself?

I am using Abraham and Freud and Ferenczi and Klein as starting points for arguing that physiology and phantasy are coeval in states of depression. Moreover, I want to stress that trends of sadism and hate are a crucial part of what is entangled and trafficked psychosomatically. Here again Abraham (1924) is helpful: he notes the sadistic character of many enteric impulses. The ambivalence (love and hate) that underwrites melancholia draws on the sadistic ways in which objects are orally and anally addressed: "In the biting stage of the oral phase the individual incorporates the object in himself and in so doing destroys it. . . . As soon as the child is attracted by an object, it is liable, indeed bound, to attempt its destruction" (Abraham 1924, 451). The loss of an object is also hostile phantastically and enterically: "The removal or loss of an object can be regarded by the unconscious either as a sadistic process of destruction or as an anal one of expulsion" (Abraham 1924,

428). As Klein (1935/1975) notes, there is nothing benign about incorporation: taking in an object—even a loved one—involves devouring and damaging it. Moreover, in both Abraham and Klein it is not clear whether the object is phantastic or enteric: the difference between incorporation and mastication is in doubt. As I argued in the first chapter, it is this uncertainty that has been a site of concern for Klein's critics—there is a failure to distinguish sharply between a physical stimulus and a phantasy, for example.[9] For my purposes this muddle is uniquely instructive about the psychosomatic ground of melancholic states, for it shows that the difference between an enteric action and a minded response cannot be definitively made. For it is not that the infant (or child or adolescent or adult) bites in order to express a sadism that is already extant; rather, biting is sadism uttered by the muscles of the face and jaw. Similarly, the bowel's physiological capacity for retention and expulsion are, among other things, biosadistic modes of control and destruction.

With the aid of these early psychoanalytic thinkers I want to argue that the actions of the gut are object relations—not simply the representation of object relations, or (more conservatively still) the biological bases for object relations. If, as Ferenczi and Isaacs have argued, phantasy is an archaic capacity of the substrata, then the routine actions of the gut (ingestion, metabolism, peristalsis, excretion) must be psychically alive to the consequences of loss. If, as Klein argues, phantasy is unable to be dissociated from sadism, then processes like ingestion, metabolism, peristalsis, and excretion are not simply internal matters—they are also actions that aggress on the world.[10] What I would like to explore in the next section is an example of how the gut is sometimes involved in extraordinary work of incorporation and expulsion. This is work that defies conventional interpretations that gut dysfunction is either a physiological malfunction (disease) or a secondary manifestation of disruptions to mind (functional or hypochondriacal disorder). This is also work that brings trends of sadism and hate to the fore. With this example in hand I would then like to see how analysis of such biological/hostile trends might have a bracing effect on feminist theories and politics.

Merycism

What follows is speculation, often far-fetched speculation, which the reader will consider or dismiss according to his individual predilection. It is further an attempt to follow out an idea consistently, out of curiosity to see where it will lead.
—Sigmund Freud, *Beyond the Pleasure Principle*

Merycism is the repeated regurgitation, rechewing, and reswallowing of food. The DSM-5 calls it "rumination disorder" and lists it as one of the feeding and eating disorders: "Previously swallowed food that may be partially digested is brought up into the mouth without apparent nausea, involuntary retching, or disgust. The food may be re-chewed and then ejected from the mouth or reswallowed" (APA 2013, 332). In infants the disorder can be dangerous: their failure to thrive, due to malnutrition, may eventually be fatal. What kind of hostile organ speech might this be?

There have been reports of merycism in the Anglophone medical literature throughout the twentieth century (Brockbank 1907; Cameron 1925; Franco et al. 1993; Geffen 1966; Wood and Astley 1952), and there is historical evidence of the disorder dating back to Aristotle (Parry-Jones 1994). As with melancholia, the exact form and parameters of the disorder are unclear, and they appear to have changed over time. The DSM-IV-TR (APA 2000), for example, claims that rumination disorder is rare; that it is most frequent in infants; that when found in adults it is usually accompanied by mental retardation; and that it may be more common in males. However, other studies have shown that rumination can be found in adolescents and adults who don't have developmental disabilities and that overall the prevalence in the adult population may be higher than previously thought, and some studies report many more female than male patients (although the numbers are so small in most reports that it is difficult to adduce any clear picture about gender differences). The newer DSM-5 makes no claim about gender differences and now notes that rumination disorder can occur across the lifespan. To add to this uncertainty, no one is quite sure about the mechanics of the regurgitation (what specific muscle groups are involved?): this isn't vomiting—it may be more like belching (Khan et al. 2000; O'Brien, Bruce, and Camelleri 1995; Olden 2001). Rumination disorder is also listed as a functional gastrointestinal disorder, where its mechanism remains unknown. The Rome III guidelines for the diagnosis of functional gastrointestinal disorders list criteria that are similar to those in the DSM. They also note that regurgitation is effortless, and doesn't occur either during sleep or when the infant is interacting with others. The solitary nature of the rumination makes the condition difficult to observe: "Such observations require time, patience, and stealth because rumination may cease as soon as the infant notices the observer" (Milla et al. 2006, 694).

This last observation—that rumination is highly sensitive to relationality—is key to the argument I would like to make here. While there is considerable flux in the presenting symptomology, there are two aspects of merycism in infants that appear to be fairly consistent across the empirical and clinical literature: (1) that it is a pleasurable condition, and (2) that it emerges (and then finds its remedy) within the parent-infant dyad. In 1925, the physician H. C. Cameron, for example, noted that rumination produced a sense of "beatitude" (875) in the infant:

> After taking the meal quite in the ordinary way the baby, as a rule, lies quiet for a time. Then begin certain purposive movements, by which the abdominal muscles are thrown into a series of violent contractions—head is held back, the mouth is opened, while the tongue projects a little and is curved from side to side so as to form a spoon-shaped concavity on its dorsal surface. After a varying time of persistent effort, sometimes punctuated by grunting or whimpering sounds, expressive of irritation at the failure to achieve the expected result, with each contraction of the abdominal muscles milk appears momentarily in the pharynx at the back of the mouth, as the column of fluid is pressed upwards through the stretched oesophagus. In this way the head of the column of milk may be momentarily visible many times as it attains its highest point, only to fall back again out of sight when the pressure is relaxed. Finally a successful contraction ejects a great quantity of milk forwards into the mouth. The infant lies with an expression of supreme satisfaction upon its face, sensing the regurgitated milk and subjecting it to innumerable sucking and chewing movements. (Cameron 1925, 875)

As the contemporary medical and psychiatric literatures become more concerned with the biometric assessment of rumination and with the flat cognitive-behavioral patterns of the disorder, this kind of detailed description has become rare. However even the DSM, now thought to be scrubbed clean of its Freudianism, notes the libidinal nature of rumination in infants: infants with the disorder "may give the impression of gaining satisfaction from the activity" (APA 2013, 332).[11] What I would like to note here is not simply that the regurgitation is pleasurable, but that it is a pleasure that runs counter to the usual function of the gut. While, as Freud famously noted, the swallowing of milk might readily be experienced as gratifying, the antiperistaltic movement of food is usually considered violent, distressing, and bitter in taste. This appears not to be the case with merycism in infants or in adults. One of the early accounts of merycism in adults notes that one ruminator had said the regurgitated material was "sweeter than honey, and had a more delightful relish" (Brockbank 1907, 424). Rumination appears to run counter to what Freud would call the instincts of self-preservation, and it finds a particular kind of sweetness in this revolt.

The relationality of rumination has a similar structure (gratifyingly unpleasurable). There is considerable focus in the clinical rumination literature on the importance of good-enough maternal care. Kevin Olden (2001), for example, states quite plainly that "children with infant rumination syndrome often have symptoms related to significant defects in bonding with their mother" (351), and the Rome III guidelines note the dyadic nature of the disorder: "The emotional and sensory deprivation that prompts rumination may occur in sick infants living in environments that prevent normal handling (e.g., neonatal intensive care units). It may also occur in otherwise healthy infants whose mothers are emotionally unconnected. Maternal behavior may appear to be neglectful or slavishly attentive, but in every case, there is no enjoyment in holding the baby" (Hyman et al. 2006, 1521). I would argue that the dyadic nature of infant-parent relationships relies on more than just parental goodwill; it is a more dynamic (and more emotionally equivocal) relationship than that. It is perhaps too easy to take from these clinical summaries that once maternal responsiveness has been restored (she takes enjoyment in holding her baby), the dyad is now trafficking primarily in benevolence. Manfred Menking et al. (1969), for example, report on the case of an "emaciated, chronically ill, dehydrated infant" (802) whose rumination was attributed to the disturbed nature of his mother's care: "She cared for her child in a passive, mechanical manner with an expressionless face and with no apparent desire to hug or kiss him. . . . Feeding was also performed mechanically, quite often too fast, and the mother paid little attention and adaptation to the infant's pace. He appeared tense and unrelaxed, crying and spitting repeatedly" (803–804). The infant's well-being improves dramatically once intrusive medical testing is stopped and more effort is made to emotionally engage him: "The nurses held the child frequently, and smiled, talked and played with him continuously during his waking hours" (804). When contained by good-enough care, the infant is the very picture of good spirits: "He became very responsive, cheerful, and charming" (804).

However, good-enough infant-caregiver relationships are not as unilaterally determined as this; they are continually in the process of mutual disjunction and repair. Edward Tronick, an accomplished infant development researcher, suggests that successful infant and caregiver pairings cycle in and out of attunement every few seconds: "Coordination, regardless of infant age during the first year, is found only about 30% or less of the time in face-to-face interactions, and the transitions from coordinated to miscoordinated states and back to coordinated states occur about once every three to five seconds. . . . [An interaction between infant and caregiver that is going well] frequently moves from affectively positive, mutually coordinated states to affectively negative, miscoordinated states and back again on a frequent basis" (Tronick 1989, 116). While outright hostility from the mother clearly bodes ill for the relationship, it is not the case that good-enough care is devoid of negativity, on either side. It would seem that affectively negative, miscoordinated states are just as crucial to infant and mother as affectively positive, mutually coordinated ones. It's not that the negative states strengthen the bond—by definition that cannot be true. Rather, it is that some attempt at recognizing and metabolizing aggression is a crucial part of the traffic that constitutes the bond. The capacity to hate and be hated is vital for psychic viability (Bion 1959; Winnicott 1949); or, to put this another way, bonding is not a synonym for liking.

That a mother might resent, dislike, fear, or hate her child (unconsciously or otherwise) is surely comprehensible to anyone versed in feminist theory or activism in recent decades.[12] What seems more curious at this point is the absence of a theory of negativity on the side of the infant. In the Menking case history, for example, the infant's hostility seems to fade away ("his emotional disposition changed remarkably"

[Menking et al. 1969, 804]), as if the boy's true nature is amiability. But what contribution does this infant make to the ruminatory relationship that has been established with his mother? Might he be generating a hostility that contributes to the traffic between his mother and himself? Perhaps the ruminating infant isn't simply the recipient of losses; perhaps his behavior isn't always compensatory or self-enclosed (what is called "self-stimulating behavior" [Hyman et al. 2006, 695] in the flat cognitive-behavioral lexicon of psychiatry); perhaps he is also attacking those closest to him ("all handling was answered with such angry and unpleasant crying that it tended to discourage and reduce his care to the essential work of feeding and cleaning" [Menking et al. 1969, 803]). What if the infant hates his mother before he hates himself? More precisely, isn't hate one of the ways in which differences between himself and his mother are made and sustained? In such circumstances, perhaps the infant's failure to thrive is not only a danger to himself but also a (sweet) aggression against her.

My hypothesis is that ruminatory behavior in the infant is a particular kind of negativity, performed by the gut, and directed at relations of care. The ruminating infant is playing fort-da with her food/parent. In *Beyond the Pleasure Principle*, a remarkable (far-fetched) account of primal aggression that underwrites many of Klein's later innovations, Freud tells the story of a favorite game played by his (seemingly well-behaved) grandson:

> This good little boy . . . had an occasional disturbing habit of taking any small objects he could get hold of and throwing them away from him into a corner, under the bed, and so on, so that hunting for his toys and picking them up was often quite a business. As he did this he gave vent to a loud, long-drawn-out "o-o-o-o," accompanied by an expression of interest and satisfaction. His mother and the writer of the present account were agreed in thinking that this was not a mere interjection but represented the German word "fort" [gone]. I eventually realized that it was a game and that the only use he made of any of his toys was to play "gone" with them. One day I made an observation which confirmed my view. The child had a wooden reel with a piece of string tied round it. It never occurred to him to pull it along the floor behind him, for instance, and play at its being a carriage. What he did was to hold the reel by the string

and very skillfully throw it over the edge of his curtained cot, so that it disappeared into it, at the same time uttering his expressive "o-o-o-o." He then pulled the reel out of the cot again by the string and hailed its reappearance with a joyful "da" [there]. This, then, was the complete game—disappearance and return. As a rule one only witnessed its first act, which was repeated untiringly as a game in itself, though there is no doubt that the greater pleasure was attached to the second act. (Freud 1920, 14–15)

This compulsion to repeat is evidence, Freud argues, of an unconscious trend ("a daemonic force" [36]) that works against the pleasure principle and athwart life itself: the death drive. He argues that the function of the game is not immediately apparent. The wooden reel no doubt represents the child's mother, but given that the child likely found the absence of his mother distressing, why would he repeat the event of his mother's departure over and again? And why would the first part (fort/gone) be particularly enjoyable? In the first instance, it seems that the child is taking an active stance in relation to a situation (his mother's departure) in which he is usually passive and often overwhelmed. However, this is not abreaction/catharsis of bad feeling, Freud argues; rather, the boy is taking revenge against his mother. This is not a game that allows the boy to express ill-will in order that he might—to use a modern expression—*move on*. Rather, the joy of the game is that he gets to revel in his aggression—to repeat and savor it. The game hints at tendencies that are beyond conscious experience and that reach beyond the goal of pleasure: these are impulses of aggression that are primal and profoundly uninterested in the good.

Rumination is a game of disappearance and return. It involves the compulsive repetition of an event that should be distasteful, an event that works against life and the capacity to thrive. My claim is not that the infant recruits the digestive system to act out the death drive, but that the death drive—already physiologically akin with the muscles and glands of the digestive system—can make itself known as readily through merycism as through a wooden reel.[13] One visceral response to systems of care, then, might be to establish a kind of acid satisfaction in attacking the relationship itself. Merycism amplifies an "organic elasticity" (Freud 1920, 36), already in the gut, that reaches out to the mother in order to refuse her or throw her away. This is not a heroic endeavor: it is bitter, and hostile, and possibly fatal. The gut devours and rejects and reincorporates; it damages and attempts repair and thrives on its revenge. Clinical interventions into cases of merycism that have become seriously disordered may be effective in reestablishing a good-enough caregiving relationship, but they do not obliterate this fundamental, infantile bitterness.

Flamboyant Melancholy

How much do I have to hate before you see me?
—Ken Corbett, "Melancholia and the Violent Regulation of Gender Variance"

As well as attaching to things that are damaging us (Berlant 2011), we are also trying to damage the things to which we are attached. I have been arguing that the politics of depression would benefit from more attention to the hostility generated by us and directed at our loved objects, ideals, and places. While there have been lucid articulations of the ways in which hostility is directed by others at certain kinds of persons (women, people of color, queers and perverts, the poor, outcasts, outliers and deviants of all kinds), the nature of our own participation in trends of sadism and hatred toward these objects—whom we love, with whom we may identify or collaborate, or to whom we may be sexually, economically, or politically attached—remains undertheorized.

There have been a number of critical accounts about the importance of negative feelings or hostility for the politics of depression, but they tend to veer away from a full engagement with the effects of aggression directed outward at a loved object. Douglas Crimp (2002), for example, has argued that AIDS politics could be expanded by a recognition that "some misery is self-inflicted" (149), and Ann Cvetkovich (2007), drawing on the lessons of the first decades of AIDS activism, suggests that negative feeling could be "a possible resource for political action rather than its antithesis" (460). Yet Crimp—in keeping with the critical preference for our aggression to turn inward—sees destructiveness primarily in terms of masochism rather than sadism. And he places clear limits on who or what our destructive impulses may target: when we turn our rage outward (militancy), it always seems to find an unambivalently bad object (i.e., the societal structures that have "blamed, belittled, excluded, derided" [146] those with HIV/AIDS, denying them adequate health care, housing, and employment). In Crimp's analysis, outwardly turned hostility stands on firm moral ground; it is aggression for the good. This is what we usually call politics. Cvetkovich anticipates this kind of critique and notes that she doesn't intend to turn depression into a positive experience; rather, she wants to keep the negativity of depression politically salient. Nonetheless, her goals of "depathologizing" (461) negative feelings, or harnessing them as sites of "community formation" (460), are affirmative ambitions; they may cohabit with "despair and exhaustion" (467), but these are political aspirations that want to keep negativity within the reach of redemption. In making depressive feelings valuable, Cvetkovich takes from them their distinctive, elemental destructiveness; in so doing, she aggresses against the very character of the bad feeling she claims to hold dear. It is this inevitability—that politics always involves hostility against the objects that we love—that I want to explore in this final section.

Judith Butler's (1997) essay on gender melancholy—a close reading of Freud's "Mourning and Melancholia"—has been widely influential on these kinds of depressive politics and on feminist, political, critical race, and clinical theories more generally (e.g., Dimen and Goldner 2002; Eng 2000; McIvor 2012). The central argument of "Melancholy Gender / Refused Identification" is that conventional genders are melancholic in structure. In the same way that lost objects are internalized in melancholia and become part of the ego, so too lost sexual attachments in infancy are incorporated into the ego and become part of gender identity. Successful femininity, for example, requires the relinquishment of the girl's same-sex attachment to the mother and the adoption and inhabitation of a heterosexual stance. Because of the constraints of homophobic cultural and familial structures, the girl's intense attachment to her mother cannot be acknowledged and thus cannot be mourned. Instead, the lost same-sex attachment is incorporated into the little girl's ego, giving her heterosexual femininity a melancholic core. What the properly feminine girl feels herself to be, Butler argues, will be haunted by a same-sex loss that has been foreclosed.

If conventional gender is in a melancholic relation to homosexual attachments, as Butler argues, it must also be in an aggressive relation to those attachments. Gender would be susceptible to melancholic (rather than simply mournful) structuration because homosexuality has been hated before it was lost.[14] Butler pays little attention to the vicissitudes of this premelancholic hatred of homosexual attachments, except to the extent that she documents a cultural milieu in which homosexual attachments are prohibited and disavowed by the workings of heteronormativity. The idea that conventional gender aggresses against the thing it unconsciously loves (rather than just yearns for it) appears to present no trouble to Butler's overall analytic frame about the ubiquity of gendered and sexual prohibitions. However, her antipathy to thinking about such trends of sadism and hate becomes significantly more troublesome when she turns to what she calls "gay melancholia" (147). Gay melancholia seems to be the enervated response, by gay and lesbian individuals/communities, to forces of homophobia emanating from the culture at large: "what the newspapers generalize as depression" (148). Drawing on a theory of abreaction that stands at odds with her reading of Freud hitherto, Butler claims that this depression "contains anger that can be translated into political expression" (147), the Names Project Quilt being her exemplar. Here grief is unspeakable not for unconscious reasons (lost primal attachments) but because of cultural (homophobic) proscriptions.

In the first instance, then, Butler's notion of gay melancholia moves her analysis, awkwardly, from an unconscious register (gender melancholy) to a behavioral-cultural register (the AIDS quilt) without investigating what this repositioning entails. In so doing Butler breaches one of the essential axioms of psychoanalytic theory—that the unconscious is another scene that is not intelligible by (or directly translatable into) the logics of behavioral, cognitive, or social structures. That is, the loss of unconscious attachments is a different kind of tragedy than, say, the unbearable effects of AIDS deaths on the US gay community in the 1980s and 1990s (one of the events that explicitly frames Butler's analysis). Similarly, foreclosure on pre-oedipal homosexual attachments differs in important ways from the cultural injunctions against forming same-sex attachments in adolescence or adulthood. By moving from an unconscious register to a behavioral-cultural register, and by focusing on anger rather than sadism, Butler narrows the character of homosexuality considerably. Whereas gender, even in its inflexible, conventionalizing forms, is cultivated by contradictions, aggressions, hypocrisies, sadness, and longing, homosexuality is more narrowly heroic. Its sadness comes not from complex traffic between inside and out, love and hate, but more sentimentally from cruel treatment by others. Gay melancholia doesn't kill the thing it loves. Rather, it is piously trapped in "life-affirming rejoinders to the dire psychic consequences of a grieving process culturally thwarted and proscribed" (148).

But shouldn't gay melancholia entail, beyond the logic of strangulated affects and public catharsis, a hatred of the object it has loved and lost? Butler nominates heterosexuality as one of the objects that may be disavowed in the constitution of gay and lesbian identities, but she moves too quickly from this claim to a politics that might try to do better in relation to the repudiations and hostilities that order social and psychic life: "We are made all the more fragile under the pressure of such rules, and all the more mobile when ambivalence and loss are given language in which to do their acting out" (150). Lee Edelman (2004) much more faithfully understands what might be at stake in a profound hostility toward heterosexuality ("the Ponzi scheme of reproductive futurism" [4]). In so arguing, he loosens the grip of an exhausting imperative directed at homosexuality to abreact, repair, and make good. Similar claims have been made by Bersani (1987, 1990). For example, the difference between his reading of Kleinian reparation and Butler's reading of Klein accentuates two very different understandings of how to proceed politically (and psychoanalytically). Butler's (1998) comments on Klein, published about the same time as the theory of gender melancholy, only sometimes register the import of aggression ("the aggression towards the one who is lost or dead indicates a primary ambivalence in relation to others" [186]). More frequently Butler contains Kleinian aggression within an inwardly turning mechanism (the superego berates the ego), and when she focuses on outwardly directed aggression, she seeks to control it within the logic of a "poignant" (186) reparation that "thwarts" or "stifles" or "refracts" (185) damage done in the world. Butler's primary concern is how hatred *protects* rather than damages the lost/loved/hated object. Bersani (1990) might note that Butler's reading is consistent with certain later developments in Kleinian theory—a tendency to undermine Klein's early and compelling accounts of the fusion of love and aggression and sex and anxiety. In this later work, Bersani argues, Klein reinforces "culture as an unceasing effort to make life whole, to repair a world attacked by desire" (22).

The political stakes of promoting this mode of reparation are plainly articulated in Butler's (2006) analysis of being verbally abused in public. This account extends her earlier reading of gender melancholy into the contentious debates about transgender and the impact of her own work (on performativity) on trans activism and experience. While attending a slam poetry event (seemingly as a member of the public) Butler is struck by the sometimes "angry," "unrelenting," and "lucid" (68) nature of the poetry being read. Some of the transgender speakers were fighting for a conventional form of gender politics, and—in one instance—reveling in aggression against Butler herself: "It is important to note that many transsexuals were firm in their support, for instance, of a binary gender system and their right and need to fit within it. Indeed, one transwoman's poem reiterated her anger at the Michigan Women's Music Festival, at various psychiatric diagnostic terms, feminism, and after saying, literally, 'fuck you' to all these institutions, adds 'fuck you Judith Butler' " (Butler 2006, 68).

Butler brings her earlier analysis of gay melancholia to bear on these trans demands: something has been lost, and the processes by which that object might be mourned have been culturally proscribed. As before, Butler underreads aggression as anger ("a rough and lucid kind of speaking" [72]); she almost exclusively fixes on the inwardly turned nature of that anger ("the self-depriving and self-lacerating address that the ego delivers to itself" [78]), and the relationality that aggression generates when brought into the public realm turns to the good ("one that will help reconstitute a new social reality" [72]). No doubt it is dizzying to find oneself cast, unexpectedly, as the wooden reel in a game of fort-da, and identifying the strength and direction of the hostility of the first part of the game (fort/fuck you) becomes difficult. Here is one hypothesis: perhaps one of the things this particular kind of trans politics enacts is aggression against the gender that it claims to love and need—a fundamental attack on the viability of gender itself. "Judith Butler" signifies gender that is to be thrown away, retrieved, and hated, thrown away, retrieved, and hated. This aggression—how to hate and be hated—is more than evidence of a "set of [political] tensions being articulated and rearticulated in relation to one another" (80); it is a destructive force, vital to both social and psychic life. "Fuck you, Judith Butler" can be read as a bid for relationality only within a humanist frame that needs negativity, in the end, to be handmaiden to forces of attachment—a frame that hopes that destructiveness is only strangulated affect, that it may eventually find its way to public articulation and so lose its heat and its capacity to injure. That reading (Butler's reading) takes revenge against revenge, it aggresses against aggression, and thereby demonstrates the very mechanism that the content of the work cannot bring itself to acknowledge—that politics are a hateful endeavor.

Verbal abuse is one thing; surely murderous violence is another. Lawrence King was fifteen years old when he was fatally shot by a fellow student (Brandon McInerney) in the computer lab of a high school in California in 2008. King was a gender-atypical adolescent; he had taken to dressing in high heels and makeup at school. There is also evidence that he had been sexually flirtatious with McInerney in the weeks leading up to his death. The murder became a national story in the United States about bullying and gender/sexual violence in schools.[15] Ken Corbett (2009a), a psychoanalyst and theorist of masculinity, mentions the King case—in passing—in an essay on feminine boys ("retribution for gender crossing can be extreme, even deadly" [356]). The essay pleads for a more expansive understanding—in the clinic and elsewhere—of femininity in boys. In particular, Corbett wants to flesh out the phantastic nature of these boys' modes of gender. These boys are usually seen as having a troubled relation to aggression; that is, according to the dictates of conventional masculinity, they are not aggressive enough: "They tumble too little. They feel too much" (362). And while it has been common to think of their femininity as the shadow of their mother (from whom they are alleged to have not separated), Corbett more often finds in these boys a melancholic incorporation of paternal figures.

In her response to Corbett's paper, Gayle Salamon returns to the case of Lawrence King in more detail. She inquires into the wrongful death suit filed by King's parents not just against McInerney (and his family) but also against the various individuals and institutions that, the suit alleges, ought to have protected King from McInerney's violence (i.e., school officials, and workers at the residential crisis center where King had been living prior to his death). In particular, the suit argued that the high school failed to put in place adequate measures at the school that would have prevented McInerney from bringing a weapon onto campus, and they failed to follow policy about bullying (when it was common knowledge that King was the target of teasing and physical abuse). More surprisingly, the suit also targets both the high school and the residential center for failing to curb King's gender transitivity. The school ought to have disciplined King for violating the uniform dress code ("There are reasons for a dress code. Reasons why you don't put young teens in with other teens when using makeup, in heels or dressing like a girl. They knew students do not respond well to others dressing that way. . . . This proximately caused Lawrence King's death" [King v. McInerney 2009, 9]). Likewise, the suit argues that the school ought to have disciplined one of the teachers who provided female clothing for King to wear, as well as the workers at the residential center who encouraged his gender identifications and provided him with "the financial resources and transportation to acquire the clothing and makeup which encouraged his fatal behavior" (King v. McInerney 2009, 12).

Salamon (2009) argues (correctly) that this suit moves some of the burden of aggressivity from the bullies and perpetrator to King's sexual and gender transgressions: "It takes a particularly perverse turn of logic to understand King as the source of the dangerous threat in the scene of violence that culminated in his murder" (377). Without question the logic of sexual/gender panic, on which the lawsuit draws, has a long, distressing history in the United States, a history that has made homosexuality or gender atypicality a form of attack against which murderous counterviolence is justified. However, this ought not to trigger a further demand that Lawrence King himself be a nonaggressing subject. In the first instance, at the behavioral level, it is clear that King was not just a lippy kid, but a sexually and sometimes physically provocative one. The lawsuit alleges that there was "pushing and shoving *between* Lawrence King and Defendant Brandon McInerney in class" (italics added; King v. McInerney 2009, 6) and that King was "verbally aggressive . . . provoked others, was very impulsive, and caused problems at school" (11). But of course, Salamon's concern is not with King's behavior, which—even without the kind of clinical access that Corbett had to the inner life of his boy patients—we could take to be the likely outcome of an adolescent faced with significant familial, emotional, and social privation. Rather, Salamon's concern is that King's sexuality, per se, is seen as dangerous ("homosexual desire in itself is a violation and a violence" [377]).

As a way of countering this phobic trope, Salamon asks us to make a conceptual dissociation: "After suffering the homophobic slurs, taunts, and threats of his classmates in silence for months, Lawrence King allegedly began to respond to those taunts and threats, not by repeating his own threats but by flirting with Brandon" (Salamon 2009, 377). Salamon wants King's sexuality to be a nonaggressive component of the events leading up to his death, and she enacts this by making a distinction between sexuality (flirting) and violence (threats). Making flirtation (a giddy game of to-and-fro) a harmless activity seems like an overly conventional move here; but, more compellingly, it is not a move that Salamon is able to sustain. Just a few lines later she notes: "Lawrence's 'hitting on' Brandon is responded to as if it were a physical threat, as if it were a literalized hitting, even though that flirtation was offered as a *disarming* response to the threats of physical violence to which Brandon had repeatedly subjected Lawrence" (italics added; Salamon 2009, 378). It seems to me that under the chronic stress of bullying adolescents are more likely to respond with dis-arming (dismembering) responses than disarming (charming) ones, although some queer kids (and King may well have been one of them) are extremely skilled at being able to do both at once. In any case, it seems that on the available data there is very little reason to figure King as sexually and aggressively quiescent. Or, at least, it seems important to consider how we would think and write and feel about the case if we could imagine King as an aggressor of some kind; and imagine that his aggression wasn't only defensive or self-directed (although likely there was plenty of that) but that it was also fierce revenge on an object that he also loved. Moreover, it seems important to consider what revenge *we* take against King when we constrain or deflect or ignore the hostility he enacted. Do we not make him smaller and less effectual? Do we not rob him of the expansiveness that a fractious existence might sometimes foment? At these moments are we not aggressing against the boy we wish we could have protected?

We might want to say King was a flamboyant boy. And, indeed, the lawsuit is more concerned with his showy gender transgressions than his sexuality. *Flamboyant* is a word used in media reports to describe King (Salamon 2009), and it is a word that Corbett (2009a) notes has been used as a weapon in psychoanalytic theory against feminine boys. It also seems to be a colorful, scorching weapon that feminine boys and men use against gender itself. Maybe Lawrence King, in heels and makeup, enacted something that feminist theories of melancholia and gender have yet to fully articulate: the inevitability of aggression against a loved object. There are myriad ways in which a fifteen-year-old gender-transgressive boy might hate the femininity (or the boy) that he also loves and cannot have, and might delight in the sweetness of that hatred. His attempts at reparation in relation to gender must also be acts of hostility. Without more information from King it is impossible to say what those configurations may have been. My point here is more general than an analysis of King himself: if we are unable to keep our conceptual focus on these intensely hostile forces, then our politics become redemptive—sadism is tamed, culture is salvaged, politics marches on to the good. For me, at least, the unacknowledged hostility necessary to enforce those political phantasies is the most bitterly depressing scene of all.

Conclusion

It is not my goal to argue for feminist theories and politics that do no harm. If, as I have argued, it is not possible to simply walk away from the antibiologism that is laced into feminist theory, similarly it is not possible to purify politics of antagonism and injury. Biology and aggression can be ignored, perhaps, but they cannot be defanged. Where does that leave us in relation to the pharmaceutical treatments of depression? It has been more common for feminist accounts of depression to take a skeptical, distancing, or apprehensive stance in relation to biochemistry and mood—to argue that depression has been reified or intensified by psychopharmaceutical engineering, for example. Against this trend, the second part of *Gut Feminism* reads for what is most animated in pharmaceutical treatments of depression: chemical transference, the adulteration of drug and placebo, condensations of harm and cure. In the next chapter, I explain the key factors by which antidepressant pharmaceuticals are thought to work in the body, and argue that a complex logic of transference (conveyance, relationality) is at play in these biochemical therapies. The final two chapters take up instances where apprehensive readings of antidepressants have been particularly acute—placebo, and the incitement of suicidal ideation in children and adolescents. I take with me into these chapters the axioms established here: that biology is more phantastic than we have anticipated and that feminist theory is bitter in ways that are difficult to handle.

PART II

ANTIDEPRESSANTS

CHAPTER 4

CHEMICAL

TRANSFERENCE

We may treat a neurotic any way we like, he always treats himself . . .
with transferences.

—Sándor Ferenczi, "Introjection and Transference"

One of the enduring concerns in the critical literature about depression
after Prozac is the claim that pills have come to play too prominent a
part in the treatment of depression. It is argued that the advent of Prozac and its sibling pharmaceuticals have pushed talk-based therapies
to one side (Bell 2005; Fonagy 2010); that these pills have minimized
the extent to which we see depression as social, historical, or economic (Cvetkovich 2012; Davis 2013; Leader 2008; Lewis 2006; N. Rose
2007; Trivelli 2014); that these medications disproportionately impact
women and the socially disenfranchised (Emmons 2010; Griggers 1997;
Ussher 2010; Zita 1998); and that the psychopharmaceutical industry
(Big Pharma) has pathologized normal sadness and overestimated the
scope and the nature of depression (Elliott 2003; Gardner 2003; Healy
2004; Horwitz and Wakefield 2007). Let Bradley Lewis stand in as exemplary of these concerns:

> One major consequence of Prozac was to support a new psychiatric
> psychopharmacologic discourse of human pain and suffering that
> has deeply conservative political ramifications. The new biopsychi-
> atry, as a way of talking about and organizing human pain, mini-
> mizes the psychological aspects of depression—personal longings,
> desires, and unfulfilled dreams—and it thoroughly erases its social

aspects—injustice, lack of opportunity, lack of social resources, neglected infrastructure, and systemic prejudices. Not only that, but the new biopsychiatry mystifies and naturalizes the scientific (and pharmaceutical) contribution to the discourse on depression, leaving alternative opinions increasingly difficult to sustain. (Lewis 2006, 133–134)

What holds these various critiques together is a sense of wariness in relation to psychopharmacology. These pills coerce, dupe, control, manage, reterritorialize, enhance, subjectify, or normalize individuals; they are part of a "neurochemical reshaping of personhood" (N. Rose 2004, 122) that ought to be viewed with caution. Only occasionally is a direct recommendation against taking antidepressants given (e.g., Breggin and Breggin 1994), but there is nonetheless a general sense in these commentaries that prescription of these drugs should be minimized, that their widespread use is harmful, and that claims about their efficacy are not to be trusted.

But what, exactly, are these commentaries positioning themselves against? What are the chemical characteristics of these antidepressants, and how important are these data for formulating a political stance about pharmaceutical treatments of depression? Even as claims are made in the critical literature about the powerful psychological and cultural effects of antidepressants, the pharmacology of these pills remains underexamined. The doxa that they regulate serotonin levels in central nervous system (CNS) synapses is usually acknowledged, but a more precise understanding of the function of the selective serotonin reuptake inhibitors (SSRIs) is often absent.[1] How SSRIs are represented, marketed, and prescribed has been more critically interesting than how they are physiologically absorbed, distributed, metabolized, and excreted. It is as if what is most political about antidepressants is their cultural dissemination rather than their biological circulation.[2]

In this chapter I reassess these critical inclinations. I begin by tracking the way in which antidepressant medications are metabolized in human bodies—taking the gut as an important biological and political reference point. Specifically, I focus on the *pharmacokinetics* of the SSRIs. Pharmacokinetics is a branch of pharmacology that investigates the course of a drug and its metabolites in the body. This is usually distinguished from *pharmacodynamics*, which investigates the effects of the drug on the body: "A convenient lay description of these terms is that pharmacokinetics describes what the body's physiology does to a drug, and pharmacodynamics describes what a drug does to the body" (DeVane 2009, 181). My goal is to suspend, temporarily, our analytic focus on the pharmacodynamics of SSRIs (i.e., the mechanisms of SSRI action in the CNS synapse and their relevance to mood) and pay attention to how these drugs are treated by the body: How are they ingested, absorbed, distributed, and excreted? What kinds of transportation and transformation does the metabolism of an antidepressant perform? Instead of attributing all the pharmacological agency to the pill (what it does, for better or for worse, to the body and mind), we can also think about a broader network of alliances in which, among other things, the body (gut) has powerful effects on how the drug works and how mind is mobilized. My ambition is to broaden the kinds of biological processes that could be the object of critical/feminist commentary on SSRI use, and thus foster different kinds of antidepressant politics. In particular I want to deflect critical attention from the CNS synapses, where the SSRIs are supposed to be most potent, and expand the so-called serotonin hypothesis of depression to include the gut, liver, and circulatory system. Put in familiar conceptual terms, I want to use pharmacokinetics to turn critical attention from the center (brain) to the periphery (gut). There has been a tendency (in both psychiatric and critical literatures) to neglect the itinerary of SSRIs through the body, and this unnecessarily minimizes the role that the viscera play in depression. To return to Bradley Lewis's concerns, I am wagering that I can use the new biopsychiatry to make the pharmaceutical data about SSRIs a *source* (rather than an obstruction) for alternative stances on the body, mind, and depression.

Pharmacokinetics

The route of administration is a major determinant of the onset and duration of a drug's pharmacological effects.
—Lindsay DeVane, "Principles of Pharmacokinetics and Pharmacodynamics"

Let me begin with a prosaic but important datum about antidepressant medications: they are administered orally. That is, they are manufactured in tablet form, and they are swallowed.[3] For any orally administered drug the gastrointestinal (GI) tract is the site at which the drug is absorbed into the body, so GI distress (nausea, delayed gastric emptying, constipation) is a commonly experienced adverse effect of pharmaceutical use. Because oral administration of drugs is so widespread, management of the gut's response to drugs has become a crucial part of pharmaceutical treatments. For example, there are numerous technologies available for controlling where in the GI tract drugs are released. Tablets can be specially coated so that they don't dissolve in the stomach (a low-pH environment) but will dissolve in the intestine (which has elevated pH); or pills can be manufactured to float on the gastric juices, thus extending their time in the stomach (Jantzen and Robinson 2002). In most cases, the gut itself is not the target of therapeutic action; the drug is being released into the body some distance from its intended site of action (Katzung 2012). The pathways from the gut to that "effect site" are often circuitous, and in the case of psychopharmaceuticals these routes are usually overlooked as important components of treatment. It is these visceral, peripheral pathways that have captured my critical interest.

A drug like an antidepressant is intended for the central nervous system. To get to the brain, however, an SSRI must first pass through the gut lumen. At this stage, a certain percentage of the SSRI is metabolized (broken down) by intestinal enzymes (DeVane 2009). Once the drug has passed though the gut mucosa, it is transported via the portal vein to the liver, where enzymes metabolize more of the drug. These initial actions of the gut mucosa and liver, which extract a significant portion of the parent drug from the body, are called first-pass clearance. From the liver, the remaining percentage of the drug moves into general (systemic) circulation in the body. However, SSRIs usually circulate in the blood bound to proteins (albumin), and this binding makes them pharmacologically inert. Only the small percentage of the drug that remains unbound is active and available for use (Ritschel and Kearns 2004). Once the drug is in systemic circulation, the brain is targeted rapidly, as are the liver, kidneys, and other organs that are well supplied with blood. Eventually (this can take anywhere from several minutes to several hours) muscle tissue, the remaining viscera, the skin, and the body's fat will also be infused with the drug (Wilkinson 2001). Fluoxetine/Prozac, for example, is very widely distributed in the body's tissues, with significant accumulations in the lungs (Hiemke and Härtter 2000).

The passage of an SSRI from systemic circulation into the brain is equally complex. The brain is protected by the blood-brain barrier, which prevents the direct transit of large molecules and potentially toxic solutes from the blood into the brain (Begley 2003). One of the ways in which the blood-brain barrier functions is simply obstructive— the cells that make up the walls of the brain's capillaries are so tightly packed together that drugs are not able to pass between these cells into brain tissue (as they would in other parts of the body). Prevented from passing *between* cells, drugs must pass *through* the cells. One of the most widely used methods for getting drugs to cross the blood-brain barrier is to make them lipid-soluble (Wilkinson 2001). Take, for example, the related compounds morphine, codeine, and heroin, which are differently lipid soluble. Morphine has relatively low uptake across the blood-brain barrier. Codeine can be created by slightly altering the chemical structure of morphine (replacement of a hydroxyl group with a methyl group). This increases lipid solubility, and brain uptake is increased tenfold. A further lipidization of codeine (addition of two acetyl groups) creates heroin, which has a thirtyfold increase in uptake into the brain (Begley 2003). Because they are small, lipophilic molecules, SSRIs readily pass across the blood-brain barrier (Brøsen and Rasmussen 1996). Once inside the brain, SSRIs are thought to increase the amount of serotonin that is available for CNS neurotransmission (by inhibiting serotonin reuptake in the synapse), and this action is thought to elevate mood. SSRIs are eliminated from the body via the kidneys, liver, or bowel.

There are many variations on this basic narrative of SSRI pharmacology. In particular, there are a number of important kinetic differences among the five most commonly studied SSRIs—citalopram/Celexa, fluoxetine/Prozac, fluvoxamine/Luvox, paroxetine/Paxil, and sertraline/Zoloft (Baumann 1996; Hiemke and Härtter 2000; Schatzberg and Nemeroff 2006). In the first instance, each of the SSRIs varies in terms of how much of the drug reaches systemic circulation after having passed through the gut lumen and the liver—this is called a drug's bioavailability.[4] The bioavailability of paroxetine/Paxil is around 50 percent, whereas fluoxetine/Prozac has a reasonably high bioavailability (70 percent), and fluvoxamine/Luvox is even higher (greater than 90 percent) (DeBattista 2012). This means that a drug like Luvox circulates in the blood in higher concentrations than a drug like Paxil. Second, the differences in bioavailability are amplified by the fact that the metabolites of the SSRIs can also have antidepressant effects (Leonard 1996). The metabolites of paroxetine/Paxil and fluvoxamine/Luvox are psychopharmacologically inactive, whereas fluoxetine/Prozac and citalopram/Celexa have metabolites that are actively antidepressant in their effects. Indeed, one of the metabolites of fluoxetine (norfluoxetine) is as potent as a serotonin reuptake inhibitor as fluoxetine itself is (Zahajszky, Rosenbaum and Tollefson 2009). For drugs like Prozac and Celexa the difference between a drug and the by-product of a drug is not clear; in these cases the boundaries between a drug effect and a side effect, between preliminary metabolism and psychoactivity, between distribution and elimination, are constantly being made and remade. These pills are not autocratic agents that operate unilaterally on body and mind; rather, they are substances that find their pharmaceutical efficacy by being trafficked, circulated, transformed, and broken down.

A third difference among these five major SSRIs is that they vary significantly in the length of their half-life (the time it takes for the concentration of a drug in the blood to fall by 50 percent). The half-life of fluvoxamine/Luvox is fifteen hours; many of the other SSRIs have half-lives around double this (twenty-five to thirty hours). Fluoxetine/Prozac, however, has a half-life of up to seventy-two hours, and norfluoxetine has a half-life of up to sixteen days (Hiemke and Härtter 2000). This means that it can take a long time for a body to free itself from the last influences of fluoxetine and norfluoxetine. So while fluvoxamine/Luvox has a high bioavailability after first-pass clearance, it is eliminated from the body much more quickly than fluoxetine/Prozac; this makes SSRI drug effects much more mediated by the body (liver, blood, enzymes) than one might initially expect. The peripheral body is not merely a transport system for an SSRI—it is a decisive part of that drug's psychological punch.

A fourth source of variation is that the relation between dosage and plasma concentrations of the drug differs significantly within the class of SSRIs: in sertraline/Zoloft and citalopram/Celexa this relation is linear (that is, the amount of drug dispersed into the body is directly proportional to the size of the dose). In fluoxetine/Prozac, fluvoxamine/

Luvox, and paroxetine/Paxil this relationship is nonlinear, meaning that increases in dose cause disproportionate increases in plasma concentrations of the drug (Hiemke and Härtter 2000). This nonlinearity is due in part to peculiarities in enzyme metabolism in the gut. A drug like paroxetine/Paxil is both the target of metabolic enzymes and an inhibitor of those same enzymes, so certain enzymes both break down paroxetine and are inhibited in this metabolic capacity by paroxetine itself. This creates an intricate network of causality within which paroxetine is both a regulator and an object of the same enzymatic event. Adding to the entanglements generated by psychoactive metabolites, then, it seems that there is no clear demarcation in some of the SSRIs between acting and being acted upon. As well as these considerable complexities in relation to the specific structure and function of each drug, there are also differences in SSRI metabolism introduced by genomic differences between individuals, by age and gender differences, by the presence of food in the stomach at time of dosing, by the time of day, and by the ongoing physiological state of an individual (e.g., hydration).

All these data point to the active role that gut metabolism plays in the pharmacological regulation of mood. It has been conventional (in both biopsychiatric texts and the critical literatures that agitate against them) to proceed as though the cerebral synapse were an antidepressant's natural or most important coalface (e.g., "the blood-brain barrier is the single most significant factor limiting drug delivery to the CNS" [Begley 2003, 85]). However, the pharmacological data clearly indicate that antidepressants work with the whole body.[5] While SSRIs like fluoxetine/Prozac were specifically designed to engage receptor sites in the CNS, their efficacy in this task is dependent on the manner by which the gut addresses them. Any pharmaceutical alleviation (or aggravation) of depressive symptomology cannot be attributed solely to effects in cortical and subcortical structures in the brain; it must also include the actions of the enzymes in the luminal wall of the small intestine and the metabolic capacities of the liver. The philosophical issue here is whether we consider such activity to be supplemental to drug action or coeval with it: is the biological periphery exterior to mind or constitutive of it? Or, to pick up the Ferenczian logic of chapter 2, is it possible that the gut metabolism of an SSRI is an antidepressant process? Rather than presuming that all antidepressant effects happen on the other side of the blood-brain barrier, we might begin to inquire into how the peripheral body is minded terrain.

Given these psychokinetic reverberations, it is difficult to trace a straight line of influence from drug to brain to mood. There is sufficient cross-traffic within a human body in relation to the ingestion, absorption, and distribution of an SSRI to make my opening account of SSRI action (mouth to stomach to small intestine to liver to body and brain to kidneys) too simple. Moreover, I have only been considering the pharmacokinetics of the preliminary stages of drug metabolism. There are many further complexities in relation to the action of an SSRI once it arrives at a CNS synapse. For example, there are seven different families of serotonin receptors in a human CNS synapse, and each family has several subtypes, making the neurotransmission of serotonin in the synapse a highly dispersed event (Szabo, Gould, and Manji 2009). In the wake of these data I am arguing that a critical or political stance that positions itself in relation to a homogeneous and domineering class of antidepressants risks considerable insensitivity to the materiality at hand.

Let me follow one exemplary feminist critique of depression to show how this can unfold. Kimberly Emmons (2010) builds a critical account of depression and gender by examining how metaphor shapes our understanding of depression. She argues that in many cases these metaphors reduce our accounts of depression to individualized and domesticated/privatized experience. In particular, the mechanical metaphors for SSRI action in the CNS synapse (e.g., "lock and key") generate, inaccurately, the perception of a simple functionality in the pharmaceutical treatment of depression.[6] Such rhetoric "reduces depression to a chemical or mechanical problem" (110). These mechanical metaphors, she argues, promote the idea that depression can be easily targeted and ameliorated by the regulation of the central nervous system: "Rather than a whole-body approach to illness, the chemical metaphor for depression is focused singularly on neurotransmitters. The pharmaceuticals developed to correct the assumed imbalances are seen as acting solely on the receptor sites of the nerve cells of the brain. Fixing the machine of the brain—even through chemical bombardment—becomes the primary object, leaving out discussions of rebuilding or fortifying the society that fostered the malfunction in the first place" (109–110). Emmons's efforts to displace biochemical reductionism are less successful than they could be. One key difficulty is that she takes psychopharmaceutical action to be kinetically diminished and diminishing. The rhetorical association of depression with chemicals can be seen as problematic only when we already presume that chemical events are rudimentary and narrowly contained within only one biological system. I have been offering some preliminary descriptions of the metabolism of SSRIs to suggest that these pharmaceuticals are much more chemically heterogeneous than this and that they do, in fact, take a whole-body approach.

Perhaps Emmons's pivot toward "society" solves the problem of too much chemistry in the rhetoric of treatments for depression (although it is not clear to me that chemical arguments and mechanical metaphors are any more problematic than, say, the rhetoric of naturalism and balance).[7] What Emmons's approach does not address is the much more troublesome conceptual problem of the search for a singular origin for depression. To my mind, the key issue with the pharmaceutical logic of "lock and key" is less that it draws on the metaphorics of machines (which, after all, can be highly variable in their makeup) than that it circumscribes a biological locality for depression: this one lock, that one key. What Emmons's reading inadvertently brings to light is an unspoken, widely shared hope (by proponents of both chemical and social hypotheses) that an identifiable origin for depression can be found, and that intervention ("rebuilding," "fortifying") at that location will restore mental well-being. In these kinds of readings, "synapse" and "society" function as monophonic names; they signify in only one note and in order to designate only one lineage for depression. Unsurprisingly, then, there seems to be no conceptual or material space other than "synapse" and "society." In the critical space that Emmons builds, our conceptual approaches to depressive landscapes are often limited to just these two choices (the synaptic or the social), leaving aside those other parts of the body that manifest depressive states (e.g., the gut) and disregarding the internal psychic topographies that color the way in which depressions materialize differently for different people, and how all those axes cross, fortify, and undo each other. The metabolism of SSRIs is interesting to me because it breaches a path across these various terrains (synapse, gut, mind, society) which we tend to treat as disjunctive or antagonistic. Attention to an antidepressant's metabolic transformations might be one way to fashion readings of mind that are biological but nonlocalized; chemical but nondeterministic; interior yet worldly. What might then be close at hand is a pharmaceutical account of what Sándor Ferenczi called materializations (see chapter 2)—viscera that have the capacity for dynamism, motivation, and mindedness.

In the end, what is most discouraging about Emmons's argument is the way in which a pharmaceutical is walled off, conceptually, from a more general mutability or mobility (see Davis 2013 for a very similar approach). Enzymatic activity, chemical absorption, hepatic metabolism become, for her purposes, politically inert. Regrettably, this places Emmons inside the very rhetorical economy she rebukes: Wouldn't Eli Lilly also want to claim that pharmaceuticals and politics occupy distinct realms? Doesn't this expulsion of biochemistry from the scene of politics significantly reduce the material spheres that feminist critique can traverse? My proposal is this: the most potent response to the claims by Big Pharma that depression is a (narrowly defined) serotonin deficiency might not be to argue that the identification of depression with a chemical is fallacious. Rather, we could show how those very chemical actions are multifaceted and peripatetic and that their association with depression might be a way of disseminating, multiplying, and mobilizing mind.

Transference

We lack a conceptual framework for this task. The heterogeneity of biochemistry is hard to tally with the heterogeneity of psychic or social space, even though the complexities of each have been expertly articulated. If the nonlinearity of pharmacokinetic, psychic, and social processes is already well established, what remains conceptually remote is a way of articulating those nonlinearities with each other. These complexities ought not map onto each other directly; such an easy correspondence would belie the entanglements at hand. Rather, such a framework would need to be able to handle how the pharmacokinetic, psychological, and social realms both align *and* dissociate, how they are antagonistically attached. The goal could not be consilience (E. O. Wilson 2011). One name for such a framework might be *transference*.

Let me begin thinking about the question of transference in an unexpected location: the formal psychiatric criteria for depression in the DSM. These criteria are heavily reliant on somatic symptomology. Rather than reading for the reductive character of such measures, I want to extract how they are connected to the rich problematic of transference. Chronic, low-level depression (dysthymia) is diagnosed when two or more of the following are present over a two-year period: poor appetite or overeating; insomnia or hypersomnia; fatigue; low self-esteem; poor concentration or difficulty in making decisions; feelings of hopelessness. There has been some debate within psychiatry about whether the symptoms of dysthymia as described in the DSM are too strongly oriented toward somatic disturbance (appetite, insomnia, fatigue). In an official guidebook for the DSM-IV-TR (2000), arguments were made that "Dysthymic Disorder may be better characterized by a wider array of cognitive and interpersonal symptoms, especially the following: generalized loss of interest or pleasure; social withdrawal; feelings of guilt; brooding about the past; subjective feelings of irritability or excessive anger; and decreased activity, effectiveness, or productivity" (First, Frances, and Pincus 2004, 203). The DSM-5, released in 2013, combined dysthymia and chronic major depressive disorder from the DSM-IV and renamed this new hybrid condition "Persistent Depressive Disorder (Dysthymia)." However, the criteria for chronic, low-level depression remained the same as in 1994 (DSM-IV); in particular, disruptions to the soma (eating, sleeping, and vitality) remain diagnostically salient. This emphasis on bodily unrest has the advantage, to my mind, of maintaining the visibility of the organic periphery (especially the gut) in psychiatry, and it links these diagnostic criteria to earlier psychodynamic principles that were oriented to the identification and treatment of transference.

Let me explain this historically. The DSM has been revised regularly since its first publication in 1952, and there has been significant reorganization (too complex to trace here in detail) of its original Freudian nosology. For example, hysterias and depressions were originally seen as categorically distinct: hysteria was a "transference neurosis" (libido is problematically cathected/transferred to an external object), whereas melancholia was a "narcissistic neurosis" (libido is withdrawn from the world into the ego).[8] So in the DSM-I (APA 1952) and DSM-II (APA 1968) depressions are categorized as psychotic disorders (too self-enclosed to be able to engage the world) and grouped with the schizophrenic conditions. This distinction disappears in the DSM-III (APA 1980), where the depressive disorders are formally separated from the schizophrenias and are no longer considered psychotic. This separation is further consolidated in the DSM-IV (APA 1994), where depressions are classified as mood disorders, and in the DSM-5 (APA 2013), where the specific classification "depressive disorders" is used. This disintegration of the difference between transferential (hysterical) and narcissistic (depressive) conditions has meant that certain somatic symptoms, originally understood as transferential (hysterical) in origin, are now to be found scattered, textually, among other conditions such as depressions, somatic disorders, and personality disorders. These somatizations are no longer contained within a few neurotic/hysterical conditions, but are spread widely across the diagnostic terrain. It seems likely that some of the somatic symptoms of Persistent Depressive Disorder have been inherited through these nosological displacements. This means that depressive symptomology, as described most recently in the DSM-5, holds the memory of the neurotic and hysterical conditions that have been explicitly removed from psychiatric diagnosis. Importantly, these somatic traces carry with them a crucial part of the classical treatment of hysteria: the management of transference. What I will endeavor to show below is that if psychiatry is currently most concerned with the biochemistry of mood (the displacements of serotonin; the transportation of SSRIs in the body), it is not therefore any less caught up with questions of transference. The transition from somatic symptomology to biological symptomology has transformed, but not diminished, the impact of transference in depressive disorders. My suspicion is that these entangled threads (soma, transference, melancholia) could be the basis for a reading of biochemical variability that would bring some suppleness to the contemporary critical analyses of antidepressant use.

First, a clear explication of transference is needed. The OED defines transference in two ways: (1) "the action or process of transferring; conveyance from one place, person, or thing to another," and (2) (translating *Übertragung* as used by Freud) "the transfer to the analyst by the patient of re-awakened and powerful emotions previously (in childhood) directed at some other person or thing and since repressed or forgotten . . . *loosely*, the emotional aspect of a patient's relationship to the analyst." It seems important to keep these two meanings of transference alive to each other: to remember that the action of conveyance (from one person or place to another) persists inside the psychoanalytic notion of transference (an ardent emotional or intersubjective encounter); and, contrariwise, to think of the carriage of a thing (e.g., the transport of a pill through a body) as an event that might call forth an intense relationality that constitutes (and breaks down) the sovereignty of persons and places. Freud's work contains both uses of transference. Sometimes (especially in the early work or in relation to dreams), transference simply means displacement. For example, in the *Interpretation of Dreams* (1900) the shift of psychic intensity from an unconscious idea to a preconscious idea is referred to as transference. His first use of transference in the more Freudian sense comes in the *Studies on Hysteria* (1895), where he describes how a patient's unconscious wish to be kissed by a man was transferred onto Freud. He calls this carriage of the phantasy from the previous man to himself a "false connection" (Freud 1895, 303), and he sees such transferences as obstacles to a successful analysis: "The patients . . . gradually learnt to realize that in these transferences on to the figure of the physician it was a question of a compulsion and an illusion which melted away with the conclusion of the analysis" (Freud 1895, 304). The Dora case brings a more complex understanding of transference to the fore. Now Freud realizes that transference is not simply a hindrance to analysis—it is a necessary part of every treatment: "Transference, which seems ordained to be the greatest obstacle to psycho-analysis, becomes its most powerful ally, if its presence can be detected each time and explained to the patient" (Freud 1905, 117). By 1905 the connections that transference reveals are no longer considered fallacious; rather, the analysis of such remote, ambiguous, or absurd links is the very medium through which treatment works.

As is now well known, this concept of transference underwent significant reformulation during the twentieth century. In a classical Freudian metapsychology, transference is a one-sided affair—the patient moves old memories onto the analyst: "[Transferences] are new editions or facsimiles of the impulses and phantasies which are aroused and made conscious during the progress of the analysis. . . . They replace some earlier person by the person of the physician" (Freud 1905, 116). The idea of countertransference (the analyst's own unconscious reaction to the patient) initiated a reassessment of the notion of the analyst as a blank and passive recipient of transferential feeling and thus established the idea that psychoanalysis, as a form of treatment, is a *relationship* (Laplanche and Pontalis 1988). Ferenczi's experiments with technique (the infamous mutual analysis, where the analyst's transference becomes part of the data that are available to be interpreted), along with important post-Freudian work on the analytic dyad (e.g., Michael Balint, Heinz Kohut, Stephen A. Mitchell), pioneered an approach to treatment in which the relationship itself (the intersubjective field of the analytic session) is primary.

These changes in technique helped to broaden the base of psychoanalytic therapy past the classical transference neuroses, facilitating treatment of conditions like melancholia and the fragmentation of personality disorders. Or, to put this another way, the reformulation of transference disrupted the distinctions, inside psychoanalytic theory, between neurotic, melancholic, and psychotic conditions. For some contemporary analysts, transference is no longer simply the revivification of a past relationship—it is a rapport specific to the analytic dyad that requires attention in its own right: "The relational approach I am advocating views the patient-analyst relationship as continually established and reestablished through ongoing mutual influence in which both patient and analyst systematically affect, and are affected by, each other. A communication process is established between patient and analyst in which influence flows in both directions" (Aron 1991, 248). Here, it is worth noting, the transference once again begins to look and feel like a kind of conveyance or trafficking.

Thomas Ogden (1994) has been one of the most articulate contemporary theorists of the transference.[9] He begins his important paper on the "analytic third" with an assertion of the profound entanglement of analyst and analysand: "I believe that it is fair to say that contemporary psychoanalytic thinking is approaching a point where one can no longer simply speak of the analyst and analysand as separate subjects who take one another as objects" (3). Ogden doesn't just dispute the classical notion of the analyst as a blank screen; he also skillfully complicates the idea of countertransference. He focuses, not on two streams of psychic events (the analyst's and the analysand's), but on the intersubjective experience (the analytic third) that the session generates. This third intersubjectivity, owned by no one, exists in a dynamic tension with the subjectivities of the analyst and analysand: "The intersubjective and individually subjective each create, negate and preserve the other" (4). The task of analysis is not to tease apart the constituent parts of this third (How much of this experience is mine? How much is yours?) but to map out the play between individual subjectivities and the intersubjectivity of the analyst-analysand that has arisen in the room: "The analytic third is not a single event experienced identically by two people; it is an unconscious, asymmetrical cocreation of analyst and analysand which has a powerful structuring influence on the analytic relationship" (Ogden 1999, 487).[10]

Analysis of the third often requires taking into account the mundane, barely perceptible events in the room that normally escape (or are intentionally ejected) from analytic consideration (e.g., the markings on an envelope on the desk; a routine change of bodily posture). Ogden's (1994) claim is that a small, peripheral object, like an envelope, is created (in one particular session) as a new object. This emergence of the envelope as an analytic object for the first time is not simply the lifting of repression (in Ogden) in relation to the object; it is due to the communal experience of the analytic third. In the first case example that Ogden gives, there is "disappointment about the absence of a feeling of being spoken to [i.e., addressed] in a way that felt personal" (9) that is being generated by Ogden and the patient. These meanings circulate unconsciously between Ogden and the patient, within Ogden and the patient, and they gather up and refashion the envelope (with its impersonal, previously unnoticed machine-franked markings). This envelope isn't simply the recipient of Ogden's phantasy, and Ogden's feelings about the envelope aren't simply facsimiles of older feelings; rather, the traffic of this particular analytic third (the desire to be addressed personally) reconfigures the envelope along with Ogden and the patient. Reflection on the ebb and flow of this experience allows Ogden to offer an interpretation to the patient (about being stifled and suffocated) that finds its mark: "Mr. L's voice became louder and fuller in a way I had not heard before" (7). In relation to a second case (where a patient turns abruptly from the couch to look directly at Ogden, in response to Ogden's change of posture in the chair) Ogden shows how both he and the patient unconsciously understood Ogden's movement in the chair to be a sign that he was dying: "It was only at the moment described above that the noise of my movement became an 'analytic object' (a carrier of intersubjectively-generated analytic meaning) that had not previously existed. My own and the patient's capacity to think as separate individuals had been co-opted by the intensity of the shared unconscious fantasy/somatic delusion in which we were both enmeshed" (15). What Ogden is documenting here is a relationality in which each participant is profoundly (and asymmetrically) permeable to the other and to their cocreated third. Many such permeabilities might coexist in an analytic session. They can be creative or enriching or limiting or destructive, and they are not eliminable from the analytic situation.[11]

With this idea of the analytic third, Ogden allows us to think in complex ways about the nature of conventional pairings like analyst-analysand, self-other, interior-world. His claim is not just that there is no analyst without an analysand (and vice versa), but also that these parties together create distributed states of mindedness that (despite having no particular location) have significant constitutive effects on the minds of analyst and analysand. This produces a chronological paradox: the minds of analyst and analysand do not preexist the thirds they create. Together these three subjectivities generate, annul, and sustain each other. This radical permeability that is constitutive, that has no origin, and that brooks no clean lines between entities is what Karen Barad (2007) might call entanglement, or what Ferenczi (1924) might have called amphimixis.

What I would like to extract from Ogden's remarkable reformulation of the analytic relationship is a strategy for how to think about the transference of an antidepressant pill in the human body. I am not trying to map the itinerary of an SSRI directly onto the multilateral structure that Ogden has evoked; I am not interested in consilience between Ogden's third and pharmacological data. Rather, I want to extract this idea of transferential permeability as a way of thinking about the metabolism of SSRIs. We have become used to thinking of an antidepressant as a sovereign entity that is carried though the body to the brain: two distinguishable entities (pill, synapse) and two distinguishable tasks (metabolism, mood regulation). Ogden provides a schema for thinking about the entanglement, rather than distinctiveness, of pill and gut and synapse. Using Ogden, I am arguing that the treatment of depression with a pill ought not be thought of as the action of distinctly central (pharmaceutical) and peripheral (metabolic) processes. Rather, each party to the treatment is implicated in the other, distributing drug effects across the entire clinical scene. In particular, the metabolism of a drug generates psychological effects throughout the body: mindedness emerges in no particular location and with no definitive disposition—it is enteric, cerebral, hepatic, sanguineous.

The pill finds its capacity to act as an antidepressant not through reductive or deterministic or unilateral action, but through the workings of transference—which is to say, through biochemical relationality. In this sense, the work of an antidepressant may not differ significantly from the work of a humoral system, a psychotherapeutic system, or a cultural system. Effective political responses to SSRI usage could take these homologies into account; rather than accentuating the differences between a pharmaceutical treatment and an interpersonal one, for example, it might be more compelling at this time to document the ways in which they each partake in displacements and transferences. It seems to me that the current psychopharmaceutical climate demands modes of interpretation that are sympathetic with, indeed intensely caught up in, the chemical bonds of transference.

Hannah Landecker's (2013) meticulous work on the history and philosophy of metabolism helps me be more specific about the nature of SSRI transference. The concept of metabolism, as it emerges in conventional form in the nineteenth and twentieth centuries, is concerned with how an organism turns food into energy: "The body was commonly regarded as analogous to a combustion engine, into which one fed fuel. Legions of experimenters studied humans and animals as though they were balance sheets, accounting for everything that went in, and everything that came out" (194). Unsurprisingly, these conventional accounts think of metabolic physiology primarily as "the problem of two" (196): organism and environment; body and food; inside and out. Landecker traces the work of a physiologist (Claude Bernard), a biochemist (Rudolph Schoenheimer), and a philosopher (Hans Jonas) who reconfigured metabolism "as a third concept" (195): each of these thinkers, in his way, envisages metabolism as a process that disrupts easy, self-evident distinctions between an organism and its world. Bernard was able to show that animals broke down sugars ingested from the environment (as conventional physiology would predict) but that they also manufactured substances like sugar. For example, a dog's liver, even when removed from the body (and thus disconnected from the mechanics of ingestion) and completely exsanguinated, was able to continue producing sugars. A substance (sugar) that was supposed to be sourced only from outside the body and converted to fuel inside the body was also being made internally by the liver. The functions of the outside are to be found internally. Or, to put this psychodynamically, the world is introjected; and, like psychological introjections, the metabolism of sugars generates a *milieu intérieur* (the ability of living creatures to sustain a stable internal state somewhat independently of flux in their environment). Landecker notes that Bernard's work on sugar "coalesced into the cornerstone of a philosophy of life, one articulated in direct opposition to the 'dualist' framework of plants generating energy and animals consuming it" (202). The certainty of two begins to break down when the location of the environment cannot be reliably said to be outside the body. A century later, and via a series of intellectual and experimental interlocutors, the philosopher Hans Jonas amplifies the dynamic nature of metabolism as understood by Bernard. For Jonas, metabolism's relation to the inside and outside is one of entanglement: "Its function is not so much to be a boundary in between organism and environment, but to produce that distinction in the first place—to produce an 'inwardness' " (216). In this critical tradition, Landecker writes, metabolism is what makes it possible for there to be an inside and an outside. There is not "a boundary between two things, but a dynamic production of there being two things at all; without metabolism there would be no need to have inside and outside, organism and environment, animal and world. In other words, there are not two entities which enter into exchange with one another, requiring a boundary to keep them distinct, but a third thing—metabolism—which produces the two-ness of organism and environment" (217).

What Landecker and Ogden provide for an analysis of SSRIs is incredibly valuable. They give us good reason to suspect (from both a physiological and a psychological perspective) that the effects of an antidepressant pill cannot be said to be independent of the somatic, environmental, and emotionally idiosyncratic landscapes that the pill transverses. What these studies of metabolism and the unconscious also have in common is an acute awareness of the variability between individuals and within an individual, and across time. Individuals vary considerably in their metabolic kinetics, and the bodily metabolism of any individual changes according to constraints like age, temperature, weight, and activity. While psychoanalysis is often thought of as a universalizing account, the power of treatment lies in the realization that unconscious formulations will modify within a session, and over longer periods of time; such formulations will repeat and double back on (and so reconstitute) themselves. These are structuralist or systemic models that find their efficacy in the particular attachments and transformations of the clinical encounter. I am using Landecker and Ogden's work with such systems to argue that a pill, rather than being an autonomous (preexisting) agent, is fabricated as an SSRI by its somatic engagements. I have been calling this transference: both the conveyance of a thing from one place to the next, and the therapeutic way in which a pill is configured by its kinetic relation to those environments/ others. If chemical transference is what produces an SSRI, any attempt to stand either for or against antidepressant treatments will always be antitransferential (i.e., reductive). Those stances presuppose (or, rather, demand) that these pills have only one disposition—exceptionally narrow, or entirely too broad. Analyses like Landecker's and Ogden's begin the task of pushing off in a new direction (the third) where we might be able to more expansively interpret our transferential scenes.

Interpretation

In one of his key case histories, Peter Kramer (1993) discusses a patient (Lucy) who was helped enormously by both psychotherapy and SSRI antidepressants.[12] Through psychotherapy, Lucy gained a certain amount of insight into the emotional principles that organized her experience of the world—she was very sensitive to rejection and could sometimes behave self-destructively. Despite the successes of the therapeutic alliance with Kramer, there were despondent parts of her emotional makeup that the psychotherapy couldn't budge. During one particularly distressing period, Kramer prescribed Prozac, and initially Lucy responded well. The drug seemed to stabilize her relations to study and to her boyfriend, rendering her more connected to the world and more psychically robust as a consequence. Then, like some other patients, she began to feel agitated, and eventually it was decided that she should discontinue Prozac. Kramer reports that he might have started Lucy on another drug, except that she actually continued to improve without further medication. That brief period on Prozac seemed to have provided Lucy with insight: "We might say that the medication acted like an interpretation in psychotherapy" (Kramer 1993, 103). Lucy was contained and reoriented by the medication in a fashion not unlike that provided by empathically based psychotherapeutic interpretation. The chemical interpretation found its mark. Indeed, for Kramer, the actions of medicating and interpreting are broadly homologous: "It is now sometimes possible to use medication to do what once only psychotherapy did—to reach into a person and alter a particular element of personality. In deciding whether to do so, the psychopharmacologist must rely on skills we ordinarily associate with psychotherapy" (Kramer 1993, 97).

It has been common to see medication and psychotherapy as antagonistically related. The political battles over the treatment of depression often crystalize as a dispute about which of these methods is the more effective (psychologically and/or economically), or which method is more open to the cultural dimensions of mental distress. In these cases a choice needs to be made: drugs or words? Jacquelyn Zita (1998), for example, sees Kramer's attunement to Prozac as a desire to replace psychotherapeutic bonds with pharmaceutical treatments: Kramer "re-interprets the role of the psychotherapist, who still continues to work through the medium of words and analysis, as best delivered in the mode of drug transference" (65). It seems to me that this isn't right; Zita has missed the kind of transferential relations that Kramer is trying to describe. He says in relation to another patient (Susan), "For the most part, in my role as psychotherapist, I acted like a medication—like Prozac—helping to mitigate my patient's sensitivity to loss" (286). For Kramer, a medication is like an interpretation, and the therapist can act like a drug. This is a claim, not for the dominion of biology and drugs over interpretation and analysis, but for the extensive affinity between words and pills, such that each might function like the other. If the successful pharmacologist needs sometimes to act like the attuned psychotherapist, and if emotional insight can be gained somatically, this implies not simply a structural similarity between these processes, but a more intimate cohabitation of the biochemical and the psychological. Intervention in one register will reorganize patterns of organization in the other not because one register determines the other but because the two are ontologically connate.

The chemical and the psychological are sufficiently kindred for Kramer (1993) that he is unable to definitively decide whether Lucy's agitation ("objectless cravings" [106]) when taking Prozac is due to a (known) adverse effect of the medication or whether the distress results from a clearer sense, now more emotionally available to her, of her longing for her mother, who had been killed when she was a child.

Here the issue is not one of siding with drugs or siding with words (or replacing one with the other), but of tracking the relation of sympathy (fellow feeling) between words and pills.[13] To put this slightly differently: interpretations are not events confined to psychological (or cognitive) encounters. Nor are they simply actions that a cognitive system might visit upon a somatic system (a simplistic kind of suggestion). Biology too can decipher, parse, and appraise. The ruminations of the gut wall, the actions of the liver as it extracts a certain percentage of an antidepressant from the blood, the accounting of amino acids at the blood-brain barrier—these are moments of the body assaying its mental needs and limitations. If Lucy can be contained and rendered more emotionally robust by a short-term course of Prozac (which is not just in her brain, but also circulating widely through her body), this is because the interpretive capacities of her biology and her psyche are akin.

These notions are supported by some mainstream empirical research into the treatment of depression. There are two fairly robust findings in this literature. First, the combination of pharmaceutical and psychotherapeutic intervention into depression seems to work better (on average) than treatment with either pharmaceuticals or psychotherapy on their own (de Jonghe et al. 2004; Keller et al. 2000; Pampallona et al. 2004; Thase et al. 1997). For example, a paper in the *New England Journal of Medicine* in 2000 documents a study of over six hundred patients who had chronic forms of major depression. Those patients who were receiving both a psychotherapeutic treatment (short-term cognitive-behavioral analysis therapy) and a pharmaceutical treatment (nefazodone/Serzone, a non-SSRI antidepressant) had a significantly better response than those who received the psychotherapy on its own or the pharmaceutical on its own: 85 percent of those in the combined treatment group showed symptomatic improvement at twelve weeks, compared with 55 percent of those taking the antidepressant alone and 52 percent of those in the psychotherapy group (Keller et al. 2000).

Let me note initially—there is a lot to be said against this study. In an editorial in the same issue of the *New England Journal of Medicine* Marcia Angell (2000) cites this study as exemplary of the overwhelming influence of money from Big Pharma in psychiatric research: of the twelve authors of this paper, only one had no ties to the company (Bristol-Myers Squibb) that made Serzone, and the information disclosing financial ties between the authors to this and other pharmaceutical companies was so extensive that it couldn't be published in the printed form of the journal and was available only online.[14] There are a number of robust political responses that could emerge from even the shortest engagement with this study: condemnation of the role of pharmaceutical money in psychiatric research; a call for greater diversity in patient populations; a critique of the shortening and narrowing of psychotherapeutic treatment to twelve weeks of cognitive-behavioral treatment. All such interventions are important. My goal here is to add in another kind of political response, one that focuses more closely on the data themselves. What can we make of the finding (replicated many times over) that pills and words amplify each other in the treatment of depression?[15] Perhaps the choice between Freud and Prozac, between talking and ingesting, is turning out to be less ideologically and biologically definitive than we were led to believe in the postwar, propharmaceutical years of the twentieth century. What if we begin with the stance that psychoanalysis and psychopharmacology are not competing ideologies of depressive malady, but different lines of attack into the same bioaffective system? Here is evidence that the pro-Freud/anti-Freud, prodrug/antidrug debates that have occupied the political field since the antipsychiatry movement of the 1960s are becoming less potent: perhaps the difference between treating a depression biochemically and treating it psychologically or culturally is less fraught than we currently suppose. This is not to argue that pills and words are in a neutral relation to each other (clearly they are not, economically or otherwise) or that pills and words fit together seamlessly as a single form of advanced treatment. Rather, it is to argue that if we begin by thinking about words and pills as intensively, asymmetrically imbricated with each other, then we end up somewhere rather curious conceptually: medications can interpret, treatments for depression need the peripheral body, words are serotonergic, incorporation is not only in a metaphorical relation to ingestion, mind is linguistically and pharmaceutically kinetic, the subjectivities of treatment are many and varied, the coalface of treatment is dispersed psychosomatically. The politics of depression need to be able to expand (rather than police) this resonance between pills and words.

The second major finding in the research literature concerns the therapeutic relationship itself: one of the most important variables in predicting successful outcome of psychotherapy for depression is not the mode of psychological treatment (e.g., cognitive therapy vs. psychodynamic therapy) but the quality of the relationship that the therapy provides (Flückiger et al. 2012; Horvath et al. 2011; Daniel Klein et al. 2003; Krupnick et al. 1996; Martin, Garske, and Davis 2000). There is a substantial body of research, dating back to the 1970s, that investigates how the alliance between patient and clinician is a key determinant in the outcomes of treatment (Horvath et al. 2011). This notion of the alliance between patient and clinician was first coined by Elizabeth Zetzel (1956) and draws directly from Freudian theories of transference. These days the alliance is very broadly defined as "an emergent quality of partnership and mutual collaboration between therapist and client" (Horvath et al. 2011, 11), and it tends to be more oriented to conscious rather than unconscious criteria. The empirical data suggest that a strong working alliance contributes in a moderate (but robust and reliable) way to good clinical results. That is, irrespective of whether the clinician is cognitively or analytically oriented, if a strong working alliance is formed between clinician and patient, we can expect a better-than-average outcome.[16]

To put this psychoanalytically: it is the transference that cures. What alleviates depression, in cognitive-behavioral or psychodynamic or any other modality, is an intervention into the patient's patterns of relationality. This intervention, as Ogden would remind us, has a constitutive effect on clinician and patient alike: both are momentarily and then perhaps chronically reconfigured by the dynamism of their working alliance. What is curative in an alliance is not the building of self-enclosed subjectivities, but the establishment of the capacity to be permeable.[17] Here we can see how research on the alliance helps explain research on combined therapies: both bodies of research are exploring how relationality is formative, and it perhaps matters less whether these are relations between two minds, between two subjectivities and an intersubjectivity, between pills and the gut, between periphery and center, between serotonin and words. What has been appealing for me in the data about SSRIs is that when these drugs work (and even when they don't), they trace a serotonergic network that traverses the body and reanimates affinities between organs, and between biological and minded states. Effectively administered, SSRIs can promote a profound, long-lasting, permeability of the organic and psychic realms: we might say that medication sometimes acts like an interpretation.

We might say that the organic and psychic share a relational (transferential) logic.

Conclusion

This chapter is an attempt to foster feeling for pharmaceuticals in feminist and critical writing on depression. The critical literatures with which I opened this chapter have been oriented in another direction: the social nature of depression and the malfeasance of the psychopharmaceutical industry. It has been difficult in this critical environment to show a conceptual interest in the vicissitudes of serotonergic systems without raising the suspicion that such an analysis is compliant with the rhetoric, economics, and politics of Big Pharma. It is not my claim that an interest in the kinetics of serotonin treatments is free from such influence. My argument has been that rather than giving the domains of biochemistry and physiology to these corporate interests as their rightful property, we could develop a curiosity about the pharmacology of mood that would recapture biology for feminist theory. If the nature of our depressions is thoroughly imbricated with pharmaceutical forms of treatment, then conceptual and political engagement with depression needs to be biochemically literate and engaged. A feminism informed about the character of chemical transferences will be better equipped to critically assay the contemporary psychopharmaceutical scene. In the two chapters that follow I show how such a stance might play out in relation to two particularly contentious aspects of the pharmaceutical treatment of depression: the role of placebo, and the emergence of suicidal ideation.

CHAPTER 5

THE BASTARD

PLACEBO

Placebo (*plah-se'bo*) [L. "I will please"]. An inactive substance or preparation, formerly given to please or gratify a patient, now also used in controlled studies to determine the efficacy of medical substances.
—Newman Dorlan, *The American Illustrated Medical Dictionary*

The adulterated placebo, the false placebo, the bastard placebo, you might call it.
—Dr. Eugene DuBois, in Harold Wolff and Eugene DuBois, "The Use of Placebos in Therapy"

Around 1951 something strange was happening to placebo. In the decades prior, doctors in Anglophone contexts had been in the habit of regularly prescribing placebos to patients in lieu of an active medication. Howard Spiro, for example, was trained as a gastroenterologist in the United States in the 1940s; he recalls the catalogs of pills that were available to physicians at the time, several pages of which advertised placebos that could be dispensed to patients: "They came in blue and yellow and green and all the rest, and doctors were perfectly happy to have them" (Harrington 1997, 237). These special preparations had names like "Tincture of Condurango" or "Fluidextract of Cimicifuga nigra" (Pepper 1945), or sometimes they were simply bread pills or vitamins or tonics (Handfield-Jones 1953). An editorial in the *British Medical Journal* in 1952 gives anecdotal evidence that around 40 percent of prescriptions in a UK general practice were for placebo. This kind of prescribing was routine and widely understood to be beneficial for patients. The editorial argues that "there is no question about the usefulness of placebos in therapeutics, nor about the fact that they have in some cases a more powerful effect than known pharmaceutical agents" (149).

Even though the prescription of placebo was widespread, it was nonetheless thought to be a disreputable practice. In a now widely cited paper from 1945, Oliver Pepper (a professor of medicine at the University of Pennsylvania) suggests that the use of placebo has been something of an open secret: "every one of us has . . . prescribed the placebo" (409), yet this practice is "not to be mentioned in polite society" (411). Coming to the defense of placebo, Pepper argues that considerable clinical acumen is needed to use it well: the substance prescribed needs to be inert (not a sedative, for example), its name should be unknown to the patient, and it should be dispensed only in fairly limited circumstances— for example, to a patient who is waiting for the results of diagnostic tests or in cases that are "hopeless, incurable" (411) before palliative sedation is given. Other physicians argued, contentiously, that placebo was particularly valuable for certain classes of patients. For example, in opinion pieces in the Lancet in 1953, Dr. A. Barham Carter, of Ashfield Hospital Middlesex, suggests valerian as the best placebo for psychoneurotic patients, and Dr. R. P. C. Handfield-Jones, a general practitioner in Gloucestershire, argues that "some patients are so unintelligent, neurotic, or inadequate as to be incurable, and life is made easier for them by a placebo" (825).

This routine clinical use of placebo underwent a significant change in the period immediately after World War II (Kaptchuk 1998; Shorter 2011). While the prescription of placebo in daily practice continued to be common, there were increasing concerns about the ethics and the scientificity of such practices: "[The placebo] was an instrument of deception with a negative moral valence" (Shorter 2011, 195). Placebo came to look more and more like quackery; it was the kind of practice, many began to argue, that a more rigorously scientific medicine ought to avoid. Dr. Eugene Dubois calls placebo "perhaps the most dishonest group of drugs that are used by doctors" (Wolff and DuBois 1946, 1718). One of the most important elements in this realignment was the emergence of clinical trials as the new "gold standard" for medical knowledge and practice. The roots of the present-day clinical trial can be traced back to the beginning of the twentieth century. Physicians in the 1920s and 1930s began testing the efficacy of popular psychiatric treatments: for example, does the removal of teeth help alleviate symptoms of depression or psychosis? Or, is benzedrine a more effective treatment for narcolepsy than ephedrine? (Shorter 2011). Key components of the modern clinical trial were being deployed in these early investigations: treatment groups were being compared to no-treatment groups, and patients were not told which treatment group they had been assigned to (the so-called single-blind protocol). Double-blind protocols (where neither patients nor clinicians are aware of who is receiving what treatment), randomization (where patients are allocated to treatment groups randomly), and institutionalization of the trial (where drugs have to undergo clinical trials in order to receive authorization to be sold as medications) would come later, in the 1960s and 1970s. What the epigraph quote from *The American Illustrated Medical Dictionary* shows is that around 1951, as the clinical trial is materializing, the placebo is operating in two registers simultaneously: as a prescribed treatment (as outlined by Pepper), and as the no-treatment wing (the control group) of a clinical trial. This confusion, where a substance both treats and does not treat, is what I will exploit in this chapter in relation to antidepressants. Let me elucidate the nature of this confusion a little further before I turn to the details of the relationship between antidepressants and placebo.

Conventionally understood, the use of placebo in clinical trials overtakes and supplants the old-fashioned prescription use of placebo. For example, Arthur and Elaine Shapiro, in their authoritative account of placebo in medical practice, argue that the demise of prescription placebo is a pivotal moment in the history of medicine.[1] In the middle of the twentieth century, "very slowly, progress was made in improving clinical trial methodology: initially the use of the single-blind method and placebo controls, and finally, in the 1950s, the increasing acceptance and use of the double-blind method. The development of the double-blind procedure and other advances in clinical trial methodology were major steps toward weakening the hegemony of the placebo effect" (Shapiro and Shapiro 1997, 229). The Shapiros argue that prior to the 1950s physicians were usually dispensing placebos, knowingly or unknowingly. If these substances had any measurable effect on the patient, this was due to the powers of suggestion (the so-called placebo effect). Here, then, is another confusion that this chapter will explore: not only do placebos seem to treat and not treat at the same time, but inert placebos also seem to be able to be activated by the forces of persuasion. If an inert substance can be brought to action by suggestion, if a nondrug can become a drug under the sway of a clinician's care, authority, paternalism, or attention, is there not a muddle (for my purposes, an appealing muddle) between the actions of body and mind? Perhaps the drug and the nondrug (physiology and suggestion) are working the same ontological ground?

A third tangle now quickly appears: if the placebo is known to be a kind of sham treatment, why did it not go the way of leeches, trephination, and lobotomy? The strangest of the turns happening around 1951 is this: placebo ends up being folded inside the technology that was allegedly developing to weaken or eradicate it. These days, one of the markers of a robustly designed clinical trial is that it includes some kind of comparison between the drug and placebo: "Placebo concurrent control should be employed as the standard concurrent control, whenever operationally feasible, for evaluation of the effectiveness and safety of a new therapeutic intervention" (Chow and Liu 2008, 103). Placebo found a new home inside the standardized double-blind, randomized clinical trial (RCT): the clinical trials didn't reject placebo, they introjected it. Whether this makes the placebo parasitic on the drug, or the drug parasitic on the placebo, is one of the questions this chapter explores. In the meantime, we can note that as a defense against this dependency between drug and placebo, the Shapiros (and others) deploy a sprawling kettle logic: placebos are inert and medically ineffective; placebos are active but only in an ersatz way; placebo effects are strong and need to be eradicated.[2] Anne Harrington (2006) remarks on these three kinds of placebo (short-term and illusory; a control for clinical trials; a powerful mind-body phenomenon) and notes their incompatibility. She sees these different modes of placebo as the result of discrete historical contingencies ("unfinished business from the past" [191]), rather than (as I am suggesting) an interdependent set of responses to a single problem (mind-body mutuality).

To add to the confusion in the literature, sometimes placebos are chosen not because they are inert, but because they mimic the adverse effects of the drug under study (the so-called active placebo). For example, in studying antidepressants, a placebo may be used that mimics the anticholinergic effects of these drugs (e.g., increased heart rate, decreased sweating, decreased gastrointestinal mobility, impaired concentration). This means that patients and doctors are less likely to guess the group (drug or nondrug) to which the patient has been assigned, thus protecting the double-blind protocol of the study and minimizing the forces of suggestion. The convolutions intensify here: use a placebo that is experientially indistinguishable from the drug in order to delineate whether that drug modulates experience more effectively than the placebo. There is a kind of logical peculiarity in such clinical trials where drug and placebo need to be more or less the same in order that we can tell them apart. My concern is not that clinical trials are incoherent or fruitless endeavors, but that there hasn't been enough attention paid to how their capacity to differentiate between substances requires that those substances are intimately, mutually engaged. The mutuality of drug and placebo and the (unsuccessful) efforts to disentangle them are what this chapter examines.

One common reason for prescribing placebo is the alleviation of pain. The analgesic effects of placebo appear to be robust and strong. Louis Lasagna and his colleagues (1954), for example, describe how some postoperative patients will get as much pain relief from subcutaneous injections of saline as they will from injections of morphine. Moreover (and this is a phenomenon I will return to later in the chapter), those patients that respond well to placebo respond better to morphine (i.e., these patients get significantly more pain relief from morphine than do the patients who did not respond to the placebo). Research since 1954 continues to find that, in some people, placebos have powerful analgesic effects (Levine, Gordon, and Fields 1978; Turner et al. 1994; Enck, Benedetti, and Schedlowski 2008). My interests in this chapter lie with placebo and antidepressants. There are two things to note in this regard. First, the initial generation of antidepressant medications (the tricyclics and the monoamine oxidase inhibitors [MAOIs]) emerged during the historical period I have been focusing on here. That is, antidepressant medications and the RCT are sibling inventions of the 1950s (Healy 1997). While other medications (like analgesics) have a long clinical history that predates the RCT, antidepressants coevolved with the modern clinical trial. There is a particular intimacy, I will argue, between placebo and antidepressants in the RCT that needs to be more carefully elucidated. This intimacy frustrates any political demand that we take a stand against the use of antidepressant medications on the basis that they are pharmacologically weak (Healy 2004; Kirsch 2010), and it frustrates the complementary political demand (from, say, a pharmaceutical representative) that we take a stand for antidepressant medications because they are unequivocally effective. Neither side of this debate has sufficiently considered how pill and placebo are entwined historically and pharmacologically and thus how each has relied on the other, from the very beginning, for its efficacy.

Second, around 2000, the mutuality of placebo and antidepressant became something of a crisis (Harrington 2006). Following a series of articles that reviewed the RCT data on antidepressant efficacy, there was widespread publicity that antidepressants are no more effective than placebo.[3] After a decade of booming antidepressant sales, disenchantment set in (Metzl 2003); Prozac and Zoloft were wonder drugs no more. This chapter argues that what is most illuminating about the crisis is less the claims about the efficacy of antidepressants than the way in which the mutuality of drug and nondrug (physiology and suggestion) became visible, briefly, in ways that people often found alarming, infuriating, or disenchanting, but hardly ever intriguing (wondrous). I would like to identify, and intensify, the messiness of the antidepressant–placebo relationship in order to further the arguments about transference and pharmacology that are being made in these final three chapters.

The Placebo Wars

In 2002 the journal *Prevention and Treatment* published a study that investigated the efficacy of antidepressant medication. The authors analyzed published and unpublished data submitted to the US Food and Drug Administration (FDA) by pharmaceutical companies seeking approval for new antidepressant drugs. The study collated data from forty-seven clinical trials in the period 1987 to 1999. These trials had tested the efficacy of six of the most widely prescribed antidepressants: fluoxetine/Prozac, paroxetine/Paxil, sertraline/Zoloft, venlafaxine/Effexor, nefazodone/Serzone, and citalopram/Celexa. A meta-analysis of the data generated some surprising conclusions: 80 percent of the improvement in patients taking the drugs was also seen in patients in placebo control groups; and the mean difference in improvement between patients taking the drug and those in the placebo group was only a few points on the Hamilton Depression Rating Scale.[4] There appeared to be very little difference, for participants in these clinical trials, between taking an antidepressant and taking a placebo. The authors concluded that the potency of these antidepressants was "small and of questionable clinical value" (Kirsch et al. 2002, n.p.).

This 2002 study followed on from an earlier paper, published in the same journal by the same lead author, that reached similar conclusions (Kirsch and Sapirstein 1998). In 1998 a meta-analysis of nineteen clinical trials indicated that placebo response accounted for about 75 percent of the response to active antidepressant medication. This pattern was evident with selective serotonin reuptake inhibitors (SSRIs), MAOIs, tricyclics, and other medications (like lithium) that have been prescribed for depression. The authors concluded that "the placebo component of the response to [antidepressant] medication is considerably greater than the pharmacological effect" (n.p.).

The peer response to these studies was heated. For example, Donald Klein (a dominant figure in US psychiatry and psychopharmacology and a leading contributor to the DSM-III) was strongly critical of the 1998 study. He argued that Irving Kirsch and Guy Sapirstein's sampling techniques were unrepresentative and their statistical analyses of the data inadequate. He concluded, rather forcefully, that their results were derived from "a minuscule group of unrepresentative, inconsistently and erroneously selected articles arbitrarily analyzed by an obscure, misleading effect size" (Donald Klein 1998, n.p.). Jamie Horder, Paul Matthews, and Robert Waldmann (2011) argue that a similar study by Kirsch in 2008 is also "a seriously flawed analysis which draws misleading conclusions on the basis of unusual and potentially biased statistical techniques" (1278).[5] Other colleagues were passionately supportive of Kirsch's efforts. David Antonuccio, David Burns, and William Danton (2002), for example, exclaimed that "the results of [Kirsch et al.'s] analysis are stunning and offer potential for a paradigm shift in the way we view the efficacy of antidepressant medications" (n.p.). They conclude that psychotherapy, rather than a prescription, ought to be the first line of treatment for depression. Since 2000 the Kirsch placebo studies have also become part of the citational background of critical and feminist work that is suspicious of pharmacological treatments of depression (Cvetkovich 2012; Davis 2013; Emmons 2010; Ussher 2010). Not usually analyzed in any detail by these authors, the Kirsch studies now function silently in these texts as confirmation that something is amiss, indeed seriously compromised, in the pharmaceutical treatment of depression.

The Kirsch placebo studies have been a key source of data in the political battles over the use of new-generation antidepressants. Since 1998 there has been ongoing debate about whether these medications can exacerbate or perhaps even incite suicidal ideation (Healy and Whitaker 2003); various regulatory bodies in North America, the United Kingdom, Europe, and Australasia have warned that antidepressants should not be used by children, teenagers, and in some cases young adults (see chapter 6); and cultural critics have argued that pharmaceutical companies are involved in calculated disease mongering (Bell 2004; Elliott 2003, 2004; Healy 1997, 2004).[6] Increasingly, debates about antidepressants are structured according to strongly bifurcated political allegiances: do you stand for or against the use of these drugs? Even empirical studies usually fall into one of two camps. On the one hand, there are those that gather data in the interest of the pharmaceutical companies, often simply for the purposes of gaining FDA approval; indeed, most clinical trials of antidepressant medication are supported by the pharmaceutical industry (Charney et al. 2002; Walsh et al. 2002). On the other hand, there are those studies that have been designed with the intention of breaking the clinical, political, and conceptual stranglehold that pharmaceuticals now have over the modeling of depression and its treatment. The Kirsch studies are clearly of the latter type. As John Salamone (2002) notes:

> It appears to me that this pattern of polemical exaggeration [in Kirsch] is taking place in the context of an evolution of the "placebo" issue from a scientific debate into a sort of political campaign. . . . Amidst all these swirling political forces, coupled with the popular press attention to this issue, I fear that some of the essential scientific points are being underestimated or lost. My fears are not at all lessened by the title of the article, which begins "The Emperor's New Drugs." This title appears to be directed more at the popular press than at the scientific community. (N.p.)

Similarly, it has been difficult to find feminist commentary that is not already committed to one side or other of the antidepressant/anti-antidepressant debate. What is most likely to underwrite the feminist commentary on antidepressants is an aversion to biological treatments of depression, and a plea for more socially or culturally oriented interventions that will blunt the political or epistemological force of Big Pharma. Even feminist work that attempts to break out of the narrow choice between cultural and biological explanations of depression—work that incorporates a comprehensive data set (biomedical, psychological, sociocultural), and that explicitly argues that neither materialist nor discursive nor intrapsychic approaches should be privileged—can still struggle to establish a thoroughgoing account of how these domains traffic with (and are constituted by) each other. For example, Jane Ussher (2010) argues that her model of depression "recognizes the materiality of somatic, psychological and social experience," yet she also claims that this materiality is "mediated by culture, language and politics" (23). Here politics is positioned as different from (perhaps critically oriented against) materiality, and it seems that agency (the action of mediating) is on the side of politics, culture, and language. Underlying this "material-discursive-intrapsychic model" (24) is a commitment to thinking of biology and pharmacology as domains that have to be enlisted by politics (or language or culture) before feminism can make use of them. That is, despite an interest in thinking outside the constraints of both positivism or constructivism, Ussher replicates one of their shared conceits: biology and politics are distinguishable (and the job falls to the commentator or researcher to adjudicate how much of each comes into play in the etiology or treatment of depression).

To put this another way: the idea that culture and language and politics might be mediated (enlivened; reorganized; modified) by a pill seems like a significantly less appealing option for Ussher. Indeed, she concludes that pharmaceutical medication might be helpful for some cases of "extreme mental turmoil," but that it is not "necessary or appropriate" (25) for the problems of everyday life. Pharmaceuticals—isolated from the vicissitudes of day-to-day living—take on the aura of exceptionality (exceptionally powerful, perhaps; exceptionally problematic, without a doubt). This segregation of pharmaceuticals (and the materiality they are said to transform) from commonplace politics is just one way in which feminist and critical work has avoided a full engagement with the mutuality that is at the heart of depressive states. In these circumstances it becomes particularly difficult to think of drug and placebo as anything other than distinct entities that (respectively) act physiologically or psychologically. The pressure to make a political choice between these modes of action becomes intense. This chapter will consider data that, when closely examined, support the notion that it is the imbrication of drug and placebo that is at the center of antidepressant action. Rather than having to nominate either the drug or the placebo as the sole author of antidepressant effects, rather than having to reject either the drug or the placebo as a clinical hoax, we may now be in a position to read for systems of efficacy in which both drug and placebo are properly, happily adulterated.

The Kirsch studies and the commentary they elicited reveal a peculiarity in the literature on antidepressants and placebo. Even though data are being generated and interpreted according to ever more bifurcated political and economic interests, there is at the same time an overwhelming amount of interpretive and empirical noise. The conclusions drawn from the data, the suggestions for further research, the criticisms about methodology and design, and the data themselves are extraordinarily heterogeneous. The variables that ought to characterize the placebo response in depressed subjects seem unruly or profligate; over time they haven't settled into reliable patterns from which judicious treatments or reliable clinical designs could emerge. For example, the location of treatment (in or out of hospital), the age of the trial participants, the length of the trial, and the mode of psychological assessment all generate different patterns of placebo response; and all of these variables can change with different antidepressant medications. In addition, the route of drug administration (injection or oral), the use of brand name or generic medications, and even the color of pills influence the kind of placebo response found in clinical trials. Different kinds of depression also seem to generate different kinds of placebo response: dysthymic (persistent, low-level) depressions, for example, are more likely to respond to a placebo treatment than are major depressive disorders. It is not simply that some studies contradict one another; rather, it seems that variables align, dissociate, repeat, correlate, and proliferate in a mosaic of causal relations. Harrington (2006) has done an excellent job of identifying the historical and epistemological patterns that have shaped the emergence of placebo. In this chapter, I begin with an interest in the eccentricities that seem to overwhelm efforts (such as Harrington's) to classify and order the data. In the first instance, the empirical field seems particularly instructive when read for the waywardness of drug and placebo.

Let me give an example of the kind of unconventional psychopharmaceutical system these data can invoke. Timothy Walsh, Stuart Seidman, Robyn Sysko, and Madelyn Gould (2002) reviewed data contained in seventy-five placebo-controlled antidepressant trials for Major Depressive Disorder covering the period 1981–2000. This is a carefully planned and widely cited review of antidepressant clinical trials. The authors drew two major conclusions. The first of these was consonant with other findings in the literature: there was a large placebo response in these trials (anywhere from 10 to 50 percent of the subjects responded to placebo with clinically significant improvement). The second finding was something new: "In the last two decades, the proportion of patients responding to placebo has clearly increased, at the rate of approximately seven percent per decade, and a similar increase has occurred in the fraction of patients responding to active medication" (1844). This means that there are significantly more placebo responders in clinical trials for antidepressants these days than there were twenty years ago. Similarly, there are more people who respond well to antidepressant medications. Walsh and colleagues summarize their findings thus: placebo response is "variable, substantial, and increasing" (1845). There have been a number of responses to this study that argue that this increase in placebo response over time is an artifact of certain technical issues in the way clinical trials are conducted. For example, the kind of test that is used to measure depression, and the length of a trial, will affect the placebo rate (Bridge et al. 2007). Similarly, if there are too many different trial sites from which data are being gathered, placebo rates go up (presumably because there is less control over the kinds of patients recruited and the way their treatment is conducted) (Khan et al. 2010).

My central concern here is not to intervene into the particularities of these empirical studies but to note that there is variability in the drug-placebo relationship that seems to be very hard to control. Rather than develop statistical or theoretical models of this variability, the most common response has been to explain this variability away, thus keeping alive the expectation that one day drug response will be knowable apart from placebo and that each can be separately controlled.

One example of such attempts to manage drug-placebo variability is to argue that placebo should be minimized in clinical trials (so that reliable drug effects can be measured) but maximized in clinical practice (as it seems to contribute productively to good clinical outcomes) (e.g., Rief et al. 2009). In an attempt to manage data that strays, researchers are now advocating a return to the situation in 1951 where the placebo both treats and does not treat at the same time.

It is the intention of the Walsh review to speak to the temperamental nature of placebo and its capricious relations to medication. However, the authors' data also alert us to how medication has been working in concert with placebo. Drug effects, like placebo effect, are variable, substantial, and increasing. This mutual resonance between a drug and a placebo, and the adulterating effects it has on drug and placebo efficacy, is what occupies me here. Where the literature has become fixated with the task of separating drug response from placebo response, I will argue that there is a fundamental affinity between these events. The history of this intimate relationship is well established: every new antidepressant finds its identity in relation to placebo, and it is now an industry requirement that in order to measure medication effects accurately, we must also measure placebo. It is the argument of this chapter that these circumstances are not simply convention or sound methodological design; they are also an unformulated recognition that the response to medication and the response to placebo are parasitic on each other.

Parasitism

Most commentaries on the placebo effect note that the word *placebo* derives from the Latin *Placebo Domino in regione vivorum* (I shall please the Lord in the land of the living), the opening line of the Catholic Church's vespers for the dead. In the past it was not uncommon—in the absence of actual family or genuine mourners—for people to be hired to sing the vespers for the dead. These people came to be known as placebos. It was from this idea of fake mourning that the word *placebo* entered the medical lexicon to describe a sham treatment (Shapiro and Shapiro 1997). Let me note another, now obsolete, meaning of placebo that doesn't garner as much attention in critical commentaries. In the fifteenth, sixteenth, and seventeenth centuries, to call someone a placebo was to describe that person as a flatterer, a syco-phant, or a parasite.

In colloquial terms a parasite is a term of derogation: it denotes someone who lives at the expense of another, scavenging on the bodies of the dead or vulnerable (Serres 2007). In biological terms, *parasite* is an umbrella term that describes a greater variety of codependent relationships. There are organisms that require the physical support of another but do not take nutrition from them (for example, mistletoe); there are organisms that form symbiotic relations that provide nutritional benefit yet leave the other intact or unharmed (for example, gut flora); and then there are organisms that join together for nutrition, to the mutual benefit of both parties (for example, the celebrated relation between rhizobia bacteria and legumes). The clinical literature on antidepressants and placebo has been keen to dismiss any notion that (perhaps like the rhizobia and legume) drug and placebo might have coevolved and now exist in a mutually beneficial alliance. Even where the data point to precisely such a conclusion, the political imperative always seems to be to render drug and placebo distinct events. Let's follow how this happens in some research on the neurobiology of placebo.

The Neuropsychiatric Institute at UCLA has a dedicated placebo research group that takes a special interest in placebo and antidepressant medications.[7] Their goal is not simply to understand how placebo effects occur but also (and I will argue that this second ambition directly negates the first) to distinguish between a placebo response and a medication response. Specifically, they have been looking for a biological marker (a particular neurological response) that would discriminate between the placebo responder and the antidepressant responder independently of clinical or behavioral assessment. The presence of this neurological marker would enable the clinician to isolate the so-called true drug responder, and this could be done without relying on psychological tests like the HAM-D, which was originally designed to be administered by a trained clinician (Hamilton 1960) and is therefore interpersonal in nature. A key ambition underlying the search for biological markers of depression is the hope that such intersubjective, and allegedly unreliable, metrics can be circumvented. I am presuming that attempts to evade the relationality of measurement and of data are misguided (Barad 2007): a relation is not simply the smallest possible unit of analysis (Haraway 2003), it is the only possible unit of analysis.

The biological marker the UCLA group has been most interested in is cerebral perfusion (blood flow in the brain). There is now a substantial body of evidence indicating that cerebral perfusion in depressed subjects is of a different character than that in nondepressed subjects. In particular, functional brain imaging studies show significant differences in metabolic activity in the prefrontal cortex of depressed and nondepressed patients. The prefrontal cortex is thought to be principally involved in a wide range of social and interpersonal encounters (Schore 1994). There is neurological and clinical evidence that points to the centrality of prefrontal activity in facial recognition, attachment behavior, the development of the capacity for mentalization, and affect regulation: "The prefrontal cortex is the closest there is to the neural substrate of social being" (Goldberg 2001, in Fonagy et al. 2003, 434). The clinical studies also show that antidepressant medications can return the metabolic activity of the prefrontal cortex of depressed patients to patterns similar to those found in nondepressed patients (Brody et al. 2001). Rather than using brain imaging techniques, the researchers at the Neuropsychiatric Institute employed a quantitative electroencephalography (QEEG) measure that they have shown is strongly associated with prefrontal cerebral perfusion.

In 2002 a research team at the Neuropsychiatric Institute reported EEG data that distinguished between drug and placebo responders (Leuchter et al. 2002). The study collated data from fifty-one subjects who were outpatients in clinical trials at UCLA. These participants, like those in Walsh's review, met the criteria for major depressive disorder.[8] They were randomly divided into either drug or placebo conditions, and EEG measures were taken at regular intervals over an eight-week period. At the end of eight weeks of treatment, participants fell into four groups:

1. Those who have responded to placebo
2. Those who have responded to medication
3. Those who haven't responded to placebo
4. Those who haven't responded to medication

The focus of this study was not on the difference between responders and nonresponders (as it would be in a standard clinical trial to assess the efficacy of a drug); rather, it was only the group of responsive participants that came under scrutiny. Could the EEG measures differentiate between those participants who had responded to a drug and those who had responded to placebo? Even if drug responders and placebo responders seemed the same clinically, and perhaps felt the same experientially, could they be differentiated physiologically? The answer was yes. By the end of the second week and continuing through until the end of the treatment period, there was a notable and consistent difference in EEG readings between the subjects who responded to placebo and those who responded to medication. While the responsive patients all looked the same clinically (and may well have had similar HAM-D scores), they could be differentiated on the basis of specific functional changes in prefrontal cerebral blood flow.

The key contribution the authors claim for the 2002 study is that it renders visible a difference that cannot be detected with a mood scale or through clinical acumen: "This study demonstrates that although the symptomatic improvement resulting from placebo and medication treatment may be similar, the two treatments are not physiologically equivalent. Both treatments affect prefrontal brain function, but they have distinct effects and time courses" (125). The authors are hardly interested in whether there is a causal link between prefrontal perfusion and therapeutic improvement in patients. It is not a biological theory of depression or placebo that concerns them. Instead, they position their findings in relation to commercial and governmental interests. High rates of placebo response in clinical trials hinder the development of new drugs—novel antidepressants will be abandoned if they fail to show greater clinical efficacy than placebo. Because clinical trials of antidepressants often struggle to differentiate between placebo responders and drug responders, they are less and less able to provide the necessary justification for getting a new drug to market. If a reliable biological marker can be found that differentiates the placebo response from the drug response, then they have a way of more effectively isolating "true drug response" (122) in clinical trials. In this way, it is argued, the effectiveness of an antidepressant can be measured independently of placebo.

This ambition speaks to important economic and institutional issues that bear on medical researchers (Rutherford and Roose 2013). This ambition also seems to be underwritten by a conceptual commitment that sees the placebo as parasitic (in the derogatory sense) on antidepressants. The researchers presume, in other words, that a reliable epistemology of antidepressant efficacy can be established only once placebo responses have been identified and eliminated. It is conventional in a clinical trial that placebo and its disreputable siblings (suggestion, hysteria) can be—ought to be—separated from pharmacological events. It is this presumption that a richly constructed psychosomatic event (lifting of depressive symptomology) can be decomposed into constituent parts—biology or psychology, physiology or milieu, pharmacology or talk—that marks the placebo-controlled clinical trial as fundamentally Boolean. Where authors like Leuchter contend that the true drug response needs to be freed from its contaminating relation to placebo, I am arguing that drug response is at its best (most "true") when it is finely attuned to placebo factors. To put this in the form of a paradox: an antidepressant drug is most clearly itself—indeed, can only fully be itself—when it has been adulterated by placebo.

If this was all there was to the Neuropsychiatric Institute's research on placebo, it would hardly warrant critical attention. After all, this attempt to disaggregate drug-placebo affiliations in order to produce reliable knowledges about depression is empirically skillful but conceptually orthodox. However, a later study from the same research group (Hunter et al. 2006) reports data that turns these conventional ambitions inside-out. Using the very same data set, the 2006 study focuses on events during the initial one-week placebo lead-in period. The placebo lead-in phase in a clinical trial is an initial period of about one week when participants are given a placebo and taken off all other psychoactive medications. The placebo lead-in phase is single-blind: the patients are told that they are taking active medication, while the clinical staff know that they are on placebo. Those participants who improve noticeably during this phase are removed from the rest of the clinical trial (they no longer meet the symptomatic requirements of the study—they are no longer sufficiently depressed). The remaining participants with reliable and consistent depressive symptomology are then randomized into either placebo or active treatment conditions.

The 2006 study focuses on those participants who made it into the main trial and were given the active medication. The study analyzes the difference between their EEG measures at the beginning and at the end of the one-week placebo lead-in phase. What they find is surprising. The biological marker of good medication response (decreases in prefrontal perfusion, as demonstrated in 2002) emerges during the placebo lead-in phase, *before an active drug has been administered*. More specifically, during the placebo lead-in phase, the brains of future drug responders preempt the neurophysiological changes that will develop once they have been given a drug; the brains of those who will be drug nonresponders show no changes during the lead-in phase. On the basis of these data, it seems that being a good medication responder means being a good placebo responder: having a good drug response seems to go hand in hand with having a good placebo response, and likewise, the less well you respond to placebo, the less well you respond to the drug (a finding, let's remember, already established in relation to pain by Lasagna and his colleagues in 1954). Moreover, these data have been obtained from participants with moderate to severe depression, where we would expect the symptomology to be less labile, and less prone to placebo than in participants with mild or dysthymic depressions.

Hunter and colleagues struggle to make sense of their findings. Ruling out the influence of a pharmacological agent (it has yet to be administered), they suggest that "treatment-related factors" (1429) may be associated with the neurophysiological changes in the lead-in phase. They list exposure to the clinical environment, interaction with clinical staff, contact with assessment protocols, beliefs, expectations, and prior experience with antidepressant medications as probable influences on the participants' mental state. Leaving aside for the moment whether or not this easy distinction between a pharmacological agent and a psychosocial factor is viable (a placebo pill surely is both), what is missing from Hunter and colleagues' initial ruminations is why treatment-related factors impact future drug responders but not future drug nonresponders. The central puzzle of treatment response remains intact: why are there neurophysiological changes in some participants but not in others, given that in the lead-in phase they are all exposed to the same pharmacological and psychosocial agents?

The first thing to note is that the issue of why some participants respond to a placebo and some do not is not reducible to conventional personality characteristics. Extensive clinical research has failed to find any reliable demographic variable (for example, gender or class) or psychological variable (for example, personality trait or IQ) that predicts placebo responsivity (Shapiro and Shapiro 1997). There have been attempts to identify other factors that will predict placebo response (levels of patient anxiety, for example, or the quality of the physician–patient relationship), but the literature remains equivocal (Kaptchuk et al. 2008; Shapiro and Shapiro 1997). Pharmacogenetic researchers are looking for genetic markers that might differentiate drug responders from drug nonresponders (Holsboer 2008). It is my wager, however, that the differences evident in this 2006 study will not be explained by genetic variation in the participant population any more readily than they could be explained by psychological traits.[9] The problem seems to be conceptual rather than empirical: it's not that we lack sufficient data, it's that we have very few conceptual frameworks for interpreting the numbers we collate. More pharmacogenetic data, mined by sophisticated biomedical technologies, and integrated with neurological and physiological data, won't in itself generate conceptual schemata for understanding the events at hand. A more comprehensive theory of causality is needed to explain the data—one that doesn't pitch physiology against suggestion, or divide pharmacological effects from treatment-related effects, but instead understands how ingestion of pills, physiological activity, mood, and therapeutic alliance are systemically aligned. That is, the origin of placebo response (and thus the origin of drug response) is to be found not in a particular location (a gene, a trait) but dispersed across a network of psycho-genetic-institutional-pharmacological action. Antidepressant response is labile not because drugs or the RCTs that measure them are ineffective but because pharmaceuticals are constituted systemically. They are not autonomous authors of their own efficacy; rather, they call on, and in turn rely on, the actions of many other (nonautonomous, coconstituted) agencies. Hunter and colleagues' suggestion that treatment-related factors generate a placebo-inflected drug response needs to stipulate that such factors are not external to the pills (simply its environmental influences) but are an inseparable part of the pills' ontology.

The ambition in 2002 to mark a clear difference between a drug response and a placebo response is undone in 2006. Focus on the neurophysiology of medication response shows that such responses are deeply implicated in placebo events, or what Hunter and colleagues call the nonpharmacodynamic milieu. Despite their desire to separate events that are "purely nonpharmacodynamic from those that have a pharmacodynamic component" (1430), what is made clear in the 2006 study is that the active or dynamic components of an antidepressant drug are not confined within an SSRI capsule. There are no pharma-codynamic agents cut off from the nonpharmacodynamic milieu. Indeed, what makes a drug in a clinical trial pharmacodynamic and psychoactive is precisely its engagement with placebo and the treatment environment. Antidepressants are always already adulterated and parasitic on the outside. In the light of these data, it is unclear what advantages are to be had in continuing to split drug and placebo from each other.

Conclusion

Most commentaries on placebo mention a canonical paper written by Henry Beecher in 1955, just at the time when the necessity for testing drugs through randomized, placebo-controlled trials was being considered. Beecher begins the essay by quoting John Gaddum (a pharmacologist): "[Placebos] have two real functions, one of which is to distinguish pharmacological effects from the effects of suggestion, and the other is to obtain an unbiased assessment of the results of experiment" (Gaddum, in Beecher 1955, 1602). If it was the job of placebo to help us differentiate between physiology and suggestion, and if it was the job of placebo to provide unbiased assessment of drug effects in clinical trials, placebo has been an instructive failure. Fifty years after Beecher, contemporary research on placebo, like that from the UCLA Neuropsychiatric Institute, is still unable to reliably separate the biological from the mental. Indeed, as such research becomes more detailed, more technically proficient, and more statistically finessed, it seems to generate data that speak more and more to the deep sympathy between antidepressants and placebo. We seem to have been here already. In 1946 a group of physicians (from Cornell University Medical College and New York Hospital) in a discussion about the clinical uses of placebo appear to understand the complications of drugs and their placebos: "I don't like the association of . . . placebos with purity" (Dr. Henry Richardson in Wolff and DuBois 1946, 1725); "any pill, whether it be sugar or medication, is in part placebo" (Wolff and DuBois 1946, 1721). Without QEEG measures, these physicians nonetheless know that "every medicinal agent which is prescribed or administered to a patient carries with it the element of suggestion. The element of suggestion reinforces the specific action of the agent" (Wolff and

DuBois 1946, 1723). Both the UCLA researchers and the postwar physicians are pondering the same mode of psyche–soma mutuality.

Feminist and critical commentators who turn away too quickly from pharmaceutical treatments miss a number of opportunities afforded by these seventy years of data. Acknowledging the often ruthless circumstances in which antidepressants are tested, marketed, sold, and prescribed, *Gut Feminism* is nonetheless arguing for another kind of political response to the bastard nature of these drugs. Rather than calling for less compromised treatments, it seems important to explore and amplify the adulterated nature of pharmaceuticals. My claim here is that adulteration (what in previous chapters I have discussed under the rubrics of transference or amphimixis) is the engine of any treatment. Every drug needs its placebo. To argue that antidepressants are compromised by this, and therefore are weak (or treacherous), is to miss the crucial political point that every treatment requires an agent that traffics across boundaries in ways that, from the beginning, have been unstable and possibly perilous. The desire for a drug uncontaminated by placebo and the desire for treatments uncontaminated by drugs strike me as politically diminished in that they hold out for a kind of purity of action. The next chapter examines how to manage this capacity of an antidepressant to be both cure and harm.

CHAPTER 6

THE PHARMAKOLOGY

OF DEPRESSION

There is no such thing as a harmless remedy. The *pharmakon* can never be
simply beneficial.
—Jacques Derrida, *Dissemination*

On February 2, 2004, the United States Food and Drug Administration (FDA) held an advisory committee meeting to hear evidence about
whether selective serotonin reuptake inhibitor (SSRI) antidepressants
were causing suicidal ideas and behavior in children and adolescents.
More specifically, the hearing sought to examine evidence for increased
rates of suicide or suicidal ideation—not in general clinical practice,
but in the FDA-approved trials that had been conducted to test the efficacy of SSRIs in pediatric populations (Leslie et al. 2005). Held at the
Bethesda Holiday Inn, the hearing took evidence from medical experts
(biostatisticians, psychiatrists, epidemiologists) and from the public
(parents of adolescents who had killed themselves when taking SSRIs;
adolescent patients who had experienced serious side effects when taking SSRIs; mental health activists; pediatricians; legal counsel). Members of the public were given two minutes each to speak. The first of
these speakers were Dr. Irving Kirsch and Dr. David Antonuccio (both
are psychologists). By 2004 Kirsch and Antonuccio had published a
number of reports arguing that SSRIs were no more effective than placebo (see chapter 5). In their allocated two minutes they repeated their
assertion that "the therapeutic benefits of antidepressants for children
is negligible at best" (US Food and Drug Administration 2004, 80).

While children often do get better on antidepressants, they said, most of this improvement seems to be due to placebo effects. Their plea to the FDA committee was this: "In order to evaluate the cost effectiveness of antidepressant use in children, the committee must consider the benefits, as well as the risks. Clinically meaningful benefits have not been adequately demonstrated in depressed children, therefore, no extra risk is warranted. An increased risk of suicidal behavior is certainly not justified by these minimal benefits. Neither are the established increased risks of other commonly reported side effects, which include agitation, insomnia, and gastrointestinal problems" (81).

I will return to this committee, and what it recommended, later in the chapter. First of all I want to extract from Kirsch and Antonuccio's testimony a set of conceptual commitments that frame the debates about suicidality and SSRIs (and that tie this controversy to my previous argument about placebo). Kirsch and Antonuccio make two seemingly contradictory claims: (1) SSRIs are ineffective, and (2) SSRIs are risky. In order for these two statements not to seem incongruous, especially within the space of a two-minute presentation, a conceptual demarcation already has to have been made and agreed upon by Kirsch and Antonuccio and by those listening. The distinction is this: we know the difference between the therapeutic effect of a drug and the side effect of a drug. Kirsch and Antonuccio are arguing that the therapeutic effect of an SSRI (its intended action of regulating CNS serotonin and thereby lifting mood) is minimal, while the side effects (or what are more commonly in the clinical literature called adverse effects) are extensive and potentially hazardous: agitation, insomnia, gastrointestinal problems, suicidality. Around the same time, and in response to this same controversy, an editorial in the Lancet makes the same demarcation: "These drugs were both ineffective and harmful in children" (Lancet 2004, 1335). This desire for a distinction between the therapeutic effects of a drug and the harmful effects of a drug, and the antidepressant politics about suicidality that attach themselves to such a distinction, are the concerns of this chapter. Jacques Derrida's work on the pharmakon launches that analysis.

Derrida's (1981) reading of Plato's Phaedrus explores the ambivalent nature of the Greek word pharmakon, which has been variously translated as "drug," "medicine," "remedy," "poison," "recipe," or "philter" (a potion with supposedly magical properties). Derrida's interest is in how writing operates as a pharmakon in Plato's text: it (writing) is both a remedy to memory and a poisoning of memory; it both strengthens and weakens the mind. My concern in this chapter is not with specific questions about writing or memory, but with the deconstructive logic that Derrida's reading has bequeathed to us under the name of pharmakon; a logic that will help us understand one of the more controversial aspects of SSRI treatments. Put most simply, Derrida's analysis of the pharmakon brings our attention to how this word always signifies in more than one direction: it can never mean just "remedy" without also meaning "poison" and "philter." The issue is not that the word *pharmakon* is incoherent, but that it "partakes of both good and ill, of the agreeable and the disagreeable" (99). This means that—the efforts of Plato notwithstanding—it is not possible to firmly distinguish between the poisonous character of the pharmakon and its ability to heal. The efforts to find a purely beneficial remedy (or even simply a better, less harmful remedy) usually suppose that a reliable, clear distinction between poison and cure is possible.

Derrida argues that attempts to master the ambiguity of the pharmakon will likely generate a series of conventional oppositions: beneficial versus harmful, inside versus out, natural versus artificial. What such conventional logics demand is not just that remedy and harm are opposed, but also that they are radically detached from each other. That is, for something to be beneficial it has to be separate from (alien to) what is harmful; injury, impairment, or maltreatment have to be exterior to (outside of, detached from) a remedy. In such a conventional schema harm harms a remedy and must therefore be excluded from any definition or enaction of a cure: first, do no harm.

The precise character of these oppositions-that-claim-to-be-discrete is the point of curiosity for Derrida. Despite appearances, he argues, remedy and harm are not detached from each other. Rather, there are ongoing negotiations between a remedy and its harms, such that remedy is always reliant on the harms it excludes. That is, a remedy needs to eliminate harm (material, conceptual, epistemological, experimental, clinical harm) in order to claim the mantle of cure; and such acts of exclusion are incessant, indispensable, constitutive. If harm has to be placed exterior to remedy (in order that this remedy might be unequivocally beneficial), then what becomes plain is that harm is essential to the nature of remedy: "The outside [harm] is already within [cure]"

(109). This doesn't mean that a harm is somehow, secretly, restorative (and thus not really a harm at all); rather, this is a claim that damage is a necessary condition of any endeavor to heal. The pharmakon, happily trading simultaneously as remedy and poison and philter, allows Derrida to show that harm is always at the scene of remedial action. A cure is always cleaved by what is exterior to its self-identification: "The *pharmakon* is that dangerous supplement that breaks into the very thing that would have liked to do without it yet lets itself *at once* be breached, roughed up, fulfilled, and replaced" (110).

For Derrida, the oppositionality generated by conventional renderings of the pharmakon will always be deconstructable: we should be able to find how the opposition (the seeming exteriority of one term to the other) is actually an ongoing, originary intimacy. The cure is always already breached, roughed up, fulfilled, replaced by harm; natural treatments must recruit artificial techniques; the external world is part of the incorporated capsule. Barbara Johnson, the translator of "Plato's Pharmacy," says of such a critical strategy: "[It] is not an examination of [a theoretical system's] flaws or imperfections. It is not a set of criticisms designed to make the system better. It is an analysis that focuses on the grounds of that system's possibility. The critique reads backwards from what seems natural" (Johnson 1981, xv). What seems natural to Kirsch and Antonuccio, and what seems natural in most of the critical commentaries on SSRIs and suicidality, is that a drug has a therapeutic effect that can be distinguished from its adverse effects, and that good psychopharmacology will bolster a drug's therapeutic effects and minimize (exteriorize) its adverse effects. To the extent that any particular drug is unable to make this distinction clear, we know it to be a bad drug. It is these presumptions that I would like to begin to denaturalize here.

There has been a tendency to see Derrida's deconstruction of the pharmakon in the light of only two terms: poison and cure. Take, for example, Asha Persson's (2004) excellent analysis of the pharmakon-like character of HIV treatments. Antiretroviral therapies keep many patients alive, but they can also significantly disfigure or disable (e.g., lipodystrophy, which causes buildup of fat in unusual places like the back and also causes excessively sunken cheeks, creating a physiognomy reminiscent of patients in the early years of the US AIDS epidemic). These treatments "have the capacity to be beneficial and detrimental to the same person at the same time" (49). Persson offers an astute reading of patient reports of what it is like to live with these treatments ("I hate my pills. They're poison. You know, and I take them every single day. . . . I mean I love the pills, you know, like they keep me alive" [53]), and how the adverse effects of these treatments re-create the visible signification of HIV positive status on the body ("the virus which might no longer be technically detectable in the hidden interior of a person's blood paradoxically becomes detectable on the surface of their body" [52]). Nonetheless, there is a tendency in places for Persson's reading to narrow toward conventional paired possibilities: antiretroviral drugs, for example, can make "some people visible and others not" (62). At these moments the systemic or grammatological structure that is at play in Derrida's reading of the pharmakon is being restrained within an orthodox political opposition: is this visibility good or bad? This restraining political gesture is faint in Persson's argument; it is more obvious in Michael Montagne's (non-Derridean) reading of the ambiguity of the pharmakon. Montagne (1996) anticipates that eventually the tangle of remedy and cure will be undone: "Only then will society's ability to promote optimally safe and effective drug use be enhanced" (23).

The more the pharmakon is read as a stitched-together paradox (remedy + harm), the more that reading will bend toward a politics of detangling and the championing of cures that are sequestered from harms. The semiology of pharmakon is more extensive (systemic) than this; it generates poison and cure and philter and recipe and charm. My argument is that what is most pernicious about the division of the pharmakon into poison and cure is not the division into two, but the act of division itself. Such divisions, be they between two terms, or three, or five, or more, are attempts to limit a general systematicity: to cut one or more terms off from a field of entanglements or patternment (Barad 2007; Kirby 2011). One common wish in antidepressant politics is to cut psychopharmaceuticals out of treatment regimes for depression (because, it is argued, pills and serotonin are foreign to the conditions that cause depression in the first place). This is Kirsch's (2010) stated aim: "I remain convinced that antidepressant drugs are not effective treatments and that the idea that depression is a chemical imbalance in the brain is a myth" (4). Against this kind of stance, I advocate seeing antidepressants as one mode of remedy-poison in a system of depressogenic differences, alliances, partialities, and disjunctions, in which no one mode (pill, neuron, synapse, mood, psyche, subcortical pathway, chemical, economy, institution, discourse, gender) and no one function (cure/poison) can be radically detached from any other mode or function (or from the system itself) and so reign supreme. It is not my argument that SSRIs are wholly benign; nor is it my argument that they cause suicidality. Rather, I want to think of SSRI action as part of a grammatological field in which remedies are always already breached by their capacity to injure. It is not my goal, then, to stand for or against SSRIs, to constitute them as better or worse than we imagine. Instead, I am interested in how antidepressant politics might change if we knew that there was no safe harbor where SSRIs could be simply calibrated as either beneficial or damaging drugs.

It is my contention that even the very best thinking about SSRIs and suicidality in pediatric populations does not yet adequately explore the systemic (*pharmakological*) character of treatments for depression. To that end, I begin with one of the most accomplished researchers on the efficacy of psychological treatments: Peter Fonagy (who is the Freud Memorial Professor of Psychoanalysis and head of the Research Department of Clinical, Educational and Health Psychology at University College London; and chief executive of the Anna Freud Centre, London). Fonagy's work represents some of the best that contemporary psychological and psychoanalytic thinking has to offer. The ways in which he falters as he builds a theory of treatment are instructive about the conceptual difficulties that ensnare us all.

Pharmakological Treatment

In an interview about the state of the art of psychoanalysis, Peter Fonagy makes an impassioned plea for more exchange between different schools of psychological theory. Psychoanalysis is impoverished, he argues, to the extent that it is unable to engage (to the extent that it attempts to detach itself from) the treatment techniques of cognitive-behavioral therapy (CBT), the experimental methods of neuroscience, or the observational practices of qualitative research (Jurist 2010). Fonagy is perhaps the leading psychodynamic researcher in the Anglophone world today. His ambitions for psychoanalysis are far-reaching, and no doubt a little vertiginous. He is petitioning not simply for a truce in the various wars between psychological schools (phantasy versus behavior; biology versus culture; cognition versus the unconscious; drives versus affects; objective versus phenomenological), but for an active engagement among these seemingly oppositional theories about how best to treat patients. This is not an appeal for a strategic pluralism, nor does Fonagy promote an ineffectual muddle of the key elements from the major schools (cognition + the unconscious! biological and cultural interaction!). Rather, he is claiming that there are shards of psychological theory and practice that, despite their apparent historical and conceptual isolation each from the other, could be used together (for and against each other) in an effective theory of mind. There is something about the brief change-oriented methods of CBT, for example, that when injected into psychoanalytic approaches can change orthodox chronologies of treatments at their core, for the better. Likewise, the traditional psychoanalytic interest in an individual's emotional history has found a place in the most recent iterations of CBT practice, twisting that practice in new, more efficacious directions. In one way Fonagy is disputing the very antagonisms that these schools have fostered among themselves for so long: he is arguing that our theories of mind need each other, speak to each other, and in some respect are already of each other. Tired of the internecine struggles between the different schools and fed up with the narcissistic reckoning of small differences among the various denominations of psychoanalysis, Fonagy has been generating an interdisciplinary theory of mind, pathology, and treatment that is groundbreaking. An important part of that innovation lies in rethinking how different psychological theories and practices consociate—not how they seamlessly fit together, but how they might be breached, roughed up, fulfilled, or replaced. Fonagy's work has been so valuable, it seems to me, because he is powerfully rearranging how psychological knowledges could affiliate with, contest, and amplify each other.

And yet there is something in Fonagy's argument that shrinks back from certain kinds of psychological affiliation. Or at least there is something in his argument that remains attached to conventional modes of thinking about mind. The problem begins with what to do with neurological data. Fonagy is enormously optimistic about the contributions of neuroscience to new theories of mind and new modes of treatment, but at the same time he sees those contributions as tightly delimited:

"I think neuroscience as it evolves will find out more and more about how the mind works. Because the brain is the organ of the mind, we will be informed about how the mind works through neuroscience. I don't think that neuroscience will develop treatments, but it will make the psychosocial treatments more effective through greater knowledge" (Jurist 2010, 5). The equation of mind with brain ("the brain is the organ of the mind") imports a cluster of difficulties into Fonagy's argument. This is a gesture that, no matter what else it is doing, inevitably marks other (nonbrain) systems as supplementary (exterior) to mind. The peripheral and enteric nervous systems, and the endocrine, cardiovascular, skeletal, digestive, immune, respiratory, reproductive, muscular, integumentary, and excretory systems (what collectively we might want to call "the biological body") can find their place only on the margins of the mind: they are systems that may impact mind, but they are not of the mind proper (Glannon 2002). This gesture unnecessarily partitions the body into minded and unminded substance, and ties a theory of mind to all manner of problematic distinctions between reason and passion and male and female and cause and effect. In particular, this conventional idea that mind can be located in one organ undermines attempts to think the mind-body problem as a systemic puzzle. It narrows the geography of mind from a diverse, overdetermined system (an asymmetrical and asynchronous mutuality of moods-objects-institutions-neurotransmitters-hormones-cognitions-economies-affects-attachments-tears-glands-images-words-gut) to a landscape within which the brain, as sovereign, presides over psychological events. This gesture does violence to the rest of the body and to other natural and social systems; importantly, it also does violence to neurology. It removes the neuron, the synapse, the receptor, the neurotransmitter, the cortical and subcortical pathway from a more general field of reciprocity and influence, reducing them to the status of a sequestered head of state—all regulations and governance, and thus deadened to the mutuality that is the lifeblood of mind.

This dilemma is most densely enacted in Fonagy's final comments in the interview. He notes that he was one of the authors of an article in the *Lancet* that investigated the safety of SSRI antidepressants for treating depression in children and adolescents (Whittington et al. 2004). This article—a meta-analysis of published and unpublished data from randomized controlled trials (RCTs)—argued that there was support for the use of fluoxetine/Prozac in pediatric populations but that four other SSRIs (paroxetine/Paxil/Seroxat; sertraline/Zoloft/Lustral; citalopram/Celexa/Cipramil; venlafaxine/Effexor) were contraindicated.[1] These other drugs didn't just fail to show improvement over placebo in depressed children and adolescents; more alarmingly, they appeared to incite serious adverse events like agitation, aggression, impulsivity, and suicidal ideation. The *Lancet* article helped intensify regulatory moves in the United Kingdom, Europe, and the United States against the prescription of SSRI antidepressants in pediatric populations (more of which below). Fonagy refers to this study in the interview to make clear his preference for psychological treatments of depression over biological treatments. He suggests that while neuroscientific research will bring the mechanism of depressive disorders to light, biological treatments (here, the use of psychopharmaceuticals) will not be helpful in the management of depression: "When we understand the mechanism of a disorder, and this is the bottom line, when we understand the mechanism of a disorder at the level of biology, at the level of neuroscience, we will also understand that there is no way psychopharmacology will help us with those things, that the only way to alter those things will be psychological. They will be much more targeted, better targeted, but they will be psychological interventions" (Jurist 2010, 7).

There are two things (at least) happening here. On the one hand, Fonagy is trying to circumvent a certain kind of biological reductionism, in which neurological data and theories come to dominate the clinical field at the expense of other modes of treatment. On the other hand (and in a way that thwarts his stand against neurological reductionism), he is upholding a conventional notion of a biological basis for psychological disorder: for Fonagy it is neuroscientific data that form the bedrock ("bottom line") of our knowledges about and treatments for depression. It is a peculiar stance: neurology is mind's substratum, yet modulating neurological events (via a pharmaceutical) can have no useful effect on mind ("there is no way psychopharmacology will help us with those things"). What remains unclear is how minded events can be so detached from their neurological foundations and from pharmaceutical action. It seems unlikely that Fonagy thinks that the neurological, pharmaceutical, and psychological spheres are autonomous, yet what is missing in his argument is a clear sense of how these different registers of mind affiliate and disaffiliate—how they annex each other, how they bind, braid, branch, and cleave. To this end, why remove psychopharmaceuticals—those tiny packets of biochemistry and placebo and economics—from the scene of treatment? Do they not provide one vehicle for thinking about how mind is materialized (developed, expanded, soothed, damaged, or shattered) by the commerce between body and brain and world?

Fonagy's objectives are clear enough: he is trying to keep psychological treatments of depression viable in an era that has been dominated by psychopharmaceuticals (Healy 1997) and in a clinical environment that has been disfigured by enormous amounts of money from the pharmaceutical industry (Big Pharma) (Angell 2005; Petryna 2009; Petryna, Lakoff, and Kleinman 2006). These problems have become particularly acute in relation to depression since the development of SSRIs in the 1980s. Critical, political, and clinical responses to these changing circumstances remain vital, but it seems that such responses often redeploy the very neurological and pharmaceutical authority that they are attempting to displace. My key question is not are you with or against Big Pharma, as if absolute detachment or wholesome cohabitation with these forces were possible. Rather, I am interested in what readings of pharmakological intra-action we can extract from the data and theories that Big Pharma place before us.[2] If neurology, gut, mind, words, and pills are entangled—always already—then no one of them is more foundational (epistemologically or ontologically) to the problem of depression. No one of them precedes the others as a cause for depression or suicidal ideation. No one of them is the principal basis for treatment. And so no one of them can be exterior to the treatment field.

Following this logic, I would like to extend Fonagy's instincts for affiliation between psychological schools to encompass a more thoroughgoing coalition between psyche and pharmaceutical. Earlier in the same interview, Fonagy notes, "There's only one brain. There's not a CBT brain and a psychoanalytic brain, and the systems brain, and the Kleinian brain—it's one brain" (Jurist 2010, 6). This allegiance to the singularity of the brain (its exteriority to discursive contestation, and its exceptional status as the foundation of mind) undermines the careful epistemological and political work Fonagy has been undertaking in relation to psychological theory. It suggests that while theories of mind are variable, divergent, contestable, entangled, and transformable, the stuff of mind (for Fonagy, neurology) is stable and unitary. This makes the brain an isolated, exceptional organ (Singh and Rose 2006; Wilson 2011). This chapter takes up these conventions about the externality and danger of pharmaceuticals in relation to depression in children and adolescents. I am particularly interested in how adverse effects have become a key concern in the critical and clinical commentaries on antidepressant use in pediatric populations. This debate about the harms of antidepressants is an ideal site for examining how the treatment of depression might be thought more pharmakologically.

Suicidal Ideation

The Agency's scientific committee, the Committee for Medicinal Products for Human Use, concluded at its 19–22 April 2005 meeting that suicide-related behavior (suicide attempt and suicidal thoughts), and hostility (predominantly aggression, oppositional behavior and anger) were more frequently observed in clinical trials among children and adolescents treated with these antidepressants compared to those treated with placebo. The Agency's committee is therefore recommending the inclusion of strong warnings across the whole of the European Union to doctors and parents about these risks. Doctors and parents will also be advised that these products should not be used in children and adolescents except in their approved indications.
—European Medicines Agency, *European Medicines Agency Finalises Review of Antidepressants in Children/Adolescents*

[Black box warning] Antidepressants increased the risk compared to placebo of suicidal thinking and behavior (suicidality) in children, adolescents, and young adults in short-term studies of major depressive disorder (MDD) and other psychiatric disorders. Anyone considering the use of [Insert established name] or any other antidepressant in a child, adolescent, or young adult must balance this risk with the clinical need. . . . Patients of all ages who are started on antidepressant therapy should be monitored appropriately and observed closely for clinical worsening, suicidality, or unusual changes in behavior. Families and caregivers should be advised of the need for close observation and communication with the prescriber. [Insert drug name] is not approved for use in pediatric patients.
—US Food and Drug Administration, "Revisions to Product Label"

In 2007 the FDA issued an updated "black box" warning about SSRI antidepressants and suicidality in children and adolescents.[3] This warning extended advice first issued in 2004. Following the hearings at the Bethesda Holiday Inn, the FDA had issued a warning about the possible adverse effects of SSRIs in pediatric patients (those under the age of eighteen). Later the same year, the FDA had taken another step and issued a black box warning for the use of SSRIs in pediatric patients. In 2007 the FDA expanded the age group that was at risk for adverse events. They cautioned that persons up to the age of twenty-five might be particularly vulnerable to one of the antidepressants' most worrying adverse effects: suicidal thinking. Although the FDA found that there were no suicides in the trials it reviewed, it was nonetheless concerned that suicidal ideation in young patients taking SSRIs was higher than it should have been: their statistical analysis found that these drugs were associated with "a modestly increased risk of suicidality" in pediatric patients (Hammad, Laughren, and Racoosin 2006, 332).[4] The FDA advised clinicians and family members to be alert for an increase in other possible adverse effects of antidepressants: agitation, increased anxiety, panic attacks, insomnia, irritability, impulsivity, and mania. In the United Kingdom, the National Institute for Health and Clinical Excellence (NICE) had already issued a warning about SSRIs in 2003. Their subsequent guide for the treatment of depression in children and young people recommended that, in cases of moderate to severe depression, antidepressants should not be offered except in combination with concurrent psychological therapy, and the nonadult patient needs to be monitored carefully (weekly) for signs of adverse drug reactions (NICE 2005). In cases of mild depression they counseled that pharmaceuticals should not be the initial method of treatment. The Canadian Psychiatric Association and the College of Family Physicians of Canada endorsed the general orientation of both the European and United States guidelines; taking a temperate position, they argued that SSRIs are "neither a panacea nor a contraindication [for pediatric depression]. . . . When properly applied and monitored, medication treatment may be of substantial benefit to some individuals" (Garland, Kutcher, and Virani 2009, 164). Immediately following the release of these warnings, prescriptions for antidepressants to adolescents decreased 40–50 percent in the United Kingdom and 10–20 percent in the United States (Gibbons et al. 2007; Wheeler et al. 2008). By 2007 it had begun to seem that these drugs were doing more harm than good in pediatric patients. That is, the difference between a harm and a good was being entrenched, and pharmakological ambiguity was being diminished.

It is adverse effects (rather than lack of clinical efficacy) that seem to be curbing the use of SSRIs. One might not have predicted this in 1989 when Prozac was first released into the US market. At that time perhaps the most salient clinical feature of fluoxetine/Prozac was the relatively low profile of its adverse effects. As I noted in the introduction, the pre-Prozac antidepressants carried with them a number of adverse effects that had limited the kinds of patients to whom they are prescribed. What made fluoxetine/Prozac and its sibling pharmaceuticals appealing was not so much increased efficacy (they appeared to be about as efficacious, clinically, as the earlier-generation pharmaceuticals) but the apparent lack of adverse effects. This meant that SSRIs were prescribed to a much wider population base than earlier antidepressants, and for milder depressive symptoms. In particular, this meant that the antidepressant market could be massively and rapidly expanded to take on the treatment of dysthymic conditions: the so-called worried well (Bell 2005).

Soon after fluoxetine/Prozac was released into the market, serious (suicidal) adverse effects in individuals taking the drug were documented. For example, in a widely cited article, Teicher, Glod, and Cole (1990) give six short clinical vignettes of patients who fared very badly on Prozac. While any class of antidepressants can initially increase the risk for suicide (seemingly by lifting psychomotor retardation and thus making the activity necessary for a suicide attempt more likely), antidepressants had not hitherto been known, they argued, to "induce severe and persistent suicidal ideation in depressed patients free of such thoughts before treatment" (207). For example, a sixty-two-year-old woman experienced "forced obsessional suicidal thoughts" in the second week of taking fluoxetine; a thirty-nine-year-old man had "nearly constant suicidal preoccupation, violent self-destructive fantasies . . . and resignation to the inevitability of suicide" (207) in the second month of treatment with fluoxetine; a thirty-nine-year-old woman who swapped from an MAOI to fluoxetine became depressed, and as her mood worsened "she began to fantasize about purchasing a gun for the first time" (208).

In accounts like this, a strong case appears to emerge for the harmful nature of this class of pharmaceutical. However, all of the patients described in this report had complex psychiatric histories: an extensive history of treatment with antidepressants, lithium, and electroconvulsive therapy (ECT) in the first case; prior experience of serious adverse effects—passive suicidal thoughts—when taking an MAOI antidepressant in the case of the thirty-nine-year-old man; and for the thirty-nine-year-old woman, a number of conditions comorbid with her depression (borderline personality disorder and temporal lobe epilepsy). Given these circumstances, the atypicality of the drug responses seems less surprising than at first glance. If we bracket our desire to adjudicate (is this drug good or bad?), these toxic responses allow us to think the question of harm in a more disseminated manner. Given the complexity of these cases (the entrenched psychological injuries), it is impossible to locate harm in any one place: psychological harm is not just here, sealed inside this capsule; it can also be traced across a field of emotional, economic, psychic, and corporeal events. What is harmful about fluoxetine/Prozac is harmful about the psychocultural landscape in general: there can be no absolute distinction between the pill and the world, and between the remedies and injuries they each enact. To put this another way: fluoxetine/Prozac doesn't visit this scene of treatment from the outside, bringing harm as if it were a foreign plague. Rather, the ability of fluoxetine/Prozac to sometimes intensify dysphoria and distress is due to the intimacy, already in place, between pills and mood. The pill knows mood so well (for better or for worse) because they are cousins in the same psycho-material world.

My argument is not that one simply has to take the good with the bad (Liebert and Gavey 2009). Rather, I am claiming that it is not clear, at any point, whether a particular event is beneficial or harmful, and this ambiguity is the basis for any successful intervention into depressive states. For example, taking a strong antipsychiatric stand against pharmaceutical treatments *in toto* and moving one's mode of intervention away from pills (to, say, natural remedies or political activism) doesn't in any way diminish the capacity for harm. Such realignments simply distribute harm in different registers, although it may be the case that some people will find those deterritorialized (Griggers 1997), cultural (Davis 2013), or creative/spiritual (Cvetkovich 2012) harms easier to manage. The claim that certain modes of depression "won't end until there is real economic justice and a better reckoning with histories of violence" (Cvetkovich 2012, 206) still supposes, along with Big Pharma, that remedies and harms can be extricated from each other and that silver bullets (serotonin levels, social justice) can be forged. Put most forcefully, I am arguing that it is not possible to build a successful scene of antidepressive treatment that doesn't also necessarily do some (serotonergic, social) harm.

Despite the wide (often sensationalized) circulation that accounts like those in Teicher, Glod, and Cole 1990 have had, the adverse effects that most occupied users in the early days of the SSRI era were more quotidian: the standard gastrointestinal side effects of many medications that are taken orally (nausea, constipation), and the nagging dissatisfaction of sexual inhibition and delayed orgasm. Eventually the case for elevated suicide risks in adults taking SSRIs weakened; the FDA, for example, has never issued a warning about suicidality and SSRIs in relation to adult patients (Hammad, Laughren, and Racoosin 2006). However, as the number of children and adolescents who have been prescribed antidepressants has increased, there has been a return to the earliest concerns about these drugs—that they can incite atypical, frightening, and dangerous ideation.[5]

There are a number of things to note about depression in nonadult populations that bear on the issue of pharmaceutically incited suicidality. In the first instance, the criteria for depression in children and adolescents in the DSM-5 note that depression can take the form of irritability in pediatric populations: "In children and adolescents, an irritable or cranky mood may develop rather than a sad or dejected mood" (APA 2013, 163). Rather than creating something ex nihilo, SSRI pharmaceuticals may amplify the typical profile of a nonadult depression, making a bad thing worse. If we want to claim that agitation is an adverse effect of SSRIs, then we need to remember that agitation is also an adverse effect of childhood depressions—perhaps it is an adverse effect of childhood itself. The claim that a pill may intensify agitation is not the same as the claim that a pill initiates agitation; for these reasons the location (and the origin) of agitation in pediatric RCT patients is impossible to find. Second, the United Kingdom's Committee on Safety of Medicines (CSM) observes that the bodies of children and adolescents are likely to metabolize antidepressants differently from adult bodies and that environmental factors are likely to be more salient in childhood and adolescent depressions. What is environmental in the child or adolescent world is likely to be different from adult worlds, so the way in which inside and outside are calibrated will also be distinctive. In crucial ways, nonadults are more acutely attached to the traffic between self, others, and world. As one indicator of this greater reliance on the world, we can note that children and adolescents are also more likely than adults to respond to placebo in antidepressant RCTs (Bridge et al. 2009). Perhaps pediatric patients are more likely to find the structure of a clinical trial enlivening (or at least introjectable) and are thus more able to use the RCT's commerce in pills and hope to soothe agitated-depressed states. Pediatric depression seems to be a particular species of psychosomatic response: more mutable, more jagged, less autonomous from the world.

For all these reasons it seems unsurprising that pediatric depressions would respond in a capricious fashion to pharmaceutical interventions. It is in the light of this kind of system—where body and world and mood remain more intensely affiliated than many adults remember—that the suicidal ideation incited by pharmaceuticals needs to be considered. The more we think of nonadult depressions as ongoing sedimentations of affect and cognition and sociality and nerves and blood and bone, the less likely we are to see a pharmaceutical as the cause of suicidal ideation and the more likely we are to think of a pharmaceutical as a particular kind of modulator within a complex, bio-semiological system. We know already that the causal pathways between pill and mind cannot be traced in a linear fashion (see chapter 4). It's not simply that there are many pathways, and more research is needed to map their complex interactions. Rather, I have been arguing in these last three chapters that there is a mutuality between pill and mind: each gives the other form. The pill does not act directly on ideation—it has to be ingested, absorbed, transmogrified, and transported via the bloodstream to the liver; in the liver it is metabolized and then dispersed through the entire body (fat, muscles, skin, blood-brain barrier); once in the brain the SSRI arrives at the cerebral synapse and modulates the uptake of one neurotransmitter out of the scores of peptides, amino acids, and monoamines that regulate chemical traffic in the human CNS; and it is here, we are told, that ideation lies. Against this tale of the pill's "fantastic voyage," I am arguing that the SSRI derives its ideational efficacy (its capacity to cure and spoil) from its dissemination through the body and through its engagement with many different biological agencies.[6] As I argued in chapter 4, an SSRI is not a pill that dissolves once it arrives in the stomach; rather, it comes to be a psychopharmaceutical (it solidifies as an agent of treatment) only through its dispersal by the gut and liver and circulatory system and synapse. This pill—not melting but materializing—calls forth ideation not from one site (a CNS synapse) but from the body as a whole. What emerges under the rubric "suicidal ideation" in some nonadults is a kind of mindedness (no doubt a fragile, angry, hopeless kind of mindedness) that is systemically given, and that cannot be definitely known to be therapeutic or harmful. Suicidal ideation is not an isolated, extraordinary cognitive event; it is the disequilibrium of a pharmako-neuro-ideo-affect system given voice.

If the so-called cause of suicidal ideation is strewn in this manner, the nature of suicidality itself must be similarly dispersed. Indeed, it is unclear what constitutes suicidal ideation in children and adolescents in the RCT literature, beyond the meager definition provided by the FDA: "passive thoughts about wanting to be dead or active thoughts about killing oneself, not accompanied by preparatory behavior" (Bridge et al. 2007, 1684). There are no phenomenological data reported in RCTs or in the reviews of RCTs; in most cases such data have not been collected as the patients' depressions are recorded using standard psychometric scales, like the Beck Depression Inventory, that use multichoice answer formats, administered weekly. In the absence of any detailed data about suicidality, the FDA asked the manufacturers of the major SSRI antidepressants to provide "narrative summaries" (Hammad, Laughren, and Racoosin 2006, 333) of the suicide-related adverse event in their trials. These summaries were not created with reference to any narrative constraints in the patient data, but by electronic search for text strings like *suic-*, *overdos-*, *cut*, *hang*, *self harm*, *self damag-*. These summaries were then classified into five types of suicidal event by a group of pediatric suicidology experts at Columbia University. These efforts may have standardized the data across the various trials so they could be meaningfully compared statistically, but they don't bring to light any helpful data about what these depressions feel like, and how they might differ from one another. We are no closer to knowing what suicidal ideation is and what it can do. In addition to all these difficulties, suicidality is probably underreported in RCTs: in the general population suicidal thoughts appear to outnumber acts of self-harm 5:1, whereas in RCTs the reported incidence of suicidal thoughts and suicidal acts is closer to 1:1 (Committee on Safety of Medicines 2004). It may not just be that the kinds of measure used to collate data about suicidality in RCTs are somewhat blunt, the patients themselves may be recalibrating the way in which they understand the relation between suicidal ideations, feelings, actions, and words.

In these circumstances it is extremely difficult to gauge the subjective feel of these pediatric patients out of whom so much policy, political rhetoric, activism, and clinical theory is being made.[7] One of the things that could be happening in those patients who generate more suicidal ideation when taking an SSRI in a clinical trial is that the pharmaceuticals are breaching (roughing up, supplanting) the phenomenological landscape of the clinical trial. That is, this suicidality is a kind of engagement: it is torquing the highly structured system of care in which these patients finds themselves for four to sixteen weeks. Are these harmful responses? Harming responses? Remedial effects? And by what criteria could we tell the difference? David Brent and colleagues (2009) tell us something astonishing in this regard: they found that there was a rise in nonsuicidal self-harm events in their trial once they started to monitor them systematically. In the first half of the trial, self-harm events were recorded only when they were spontaneously reported by the patients (twelve- to eighteen-year-olds with moderate to severe major depressive disorder). In the second half of the trial, in response to the FDA warnings about pediatric suicidality, patients were "monitored by systematic, proactive assessment of suicidal ideation and behavior and nonsuicidal self injury" (419). That is, it became part of the clinical protocol to ask the patients about their suicidal thoughts. The detection of nonsuicidal self-injury events went from 2.2 percent to 17.6 percent once monitoring began; suicide attempts also tended to be more frequent once monitoring began (a nonsignificant increase from 3.9 percent to 6.5 percent); and both nonsuicidal self-harm and suicide attempts were reported earlier in the twelve-week trial period (at two weeks rather than five weeks). Is it more attention (more verbal inquiry, more standardized questioning, more paperwork) rather than (or in conjunction with) a higher dose of the drug that generates more suicidality?

Whether Brent and colleagues (2009) are picking up events that were otherwise not being reported, or whether the attempt to monitor adverse events actually breeds them, seems less important than understanding how these data disclose a nonlinear causal system in which drugs and rating scales and mood call on each other, incite and amplify each other. In this particular instance, neither drug nor trial protocol controls the scene. Moreover, how are we to tell, in the absence of any experiential data, whether what is being generated by this change to clinical procedure is simply self-destructive (and narrowly determined by depressive pathology) or whether it is also an attempt (albeit inadequate, dangerous, and frightening) to make a different kind of experience out of depressed states and the pharmakological (poisonous-therapeutic) feel of RCT treatments? If the use of antidepressants, in the confines of an RCT, elevates the likelihood of suicidal ideation but doesn't increase the rate of completed suicides, then perhaps this kind of ideation isn't the impairment that many researchers fear it is.[8] Perhaps the talk of suicide or self-harm, the articulation of feelings of worthlessness, the acting out of bodily self-injury are part of the benefit that the trial (and that antidepressants) can sometimes confer. Might these harms be gauges of therapeutic action? An adolescent's growing capacity to play with fire might be a sign of deteriorating mood; it might also be a sign of new curiosities (more flexibility, bravery, robustness). Or both. It would take more detailed clinical data to begin to answer such questions. I have been arguing we need a different conceptual schema in which to both procure those data and formulate responses to them.

Despite the empirical and conceptual complexities of the RCT data, it does seem clear that in some pediatric patients SSRI pharmaceuticals amplify certain kinds of ideations. But what is an ideation? The FDA and RCT literatures proceed as if ideations are primarily cognitive and linguistic (speech acts of the most traditional kind). That ideations are prone to pathological distortion only underlines that, in these trials at least, ideation is thought to be originally prudent, cogent, and sane. The *Oxford English Dictionary* locates the etymology of *ideation* in the nineteenth century, when it meant "the formation of ideas or mental images of things not present to the senses." Because the contemporary psychological literatures have been intensifying that definition (by imagining mind largely independent of senses and body and milieu), it seems helpful to think again about what might be part of the assemblage nominated in RCTs as "ideation." Here I am drawn to Silvan Tomkins's definition of an affect theory. Tomkins gives the name "theory" to the ways in which we organize our affective experiences, particularly our negative experiences. These theories are "ideo-affective organizations" (conglomerations of an affect with cognitions, perceptions, memories, actions, and bodily stances) that provide the individual with a strategy for coping with the traffic of everyday affective events (Tomkins 1963, 304). One affect (for example, shame) may dominate my response to particular kinds of event—I have lost my capacity to down-regulate and dissipate toxic feelings of humiliation. Tomkins calls such a theory monopolistic. Over time my monopolistic shame-humiliation theory may snowball and come to dominate my personality more and more. I will come to be cowed by shame, not just at this moment but at most moments of my life. I am now governed by, and attempting to manage with, a strong shame theory.

Alternatively, my shame theory might be weak—in these circumstances my shame has been socialized and down-regulated through various cognitive, ideological, and interpersonal encounters and is now sufficiently well organized that you might tend to think of me as a confident and composed person. For Tomkins, affect theories are sedimentations of the socialization of affects, cognitions about affects and bodily states (including neurological firing and facial expression). If we think of suicidal ideation as a kind of theory (rather than as isolated, inexplicable thoughts), then we have transformed its nature in important ways. Specifically, the heterogeneity of these emerging cognitive-affective organizations can now be investigated: Are they muted, angry, fearful, orderly, reasonable, fleeting, situational, sarcastic? Are there a variety of ways to passively think (theorize) about wanting to be dead? In such cases, it doesn't seem credible that one drug could incite each of these quite different ideo-affective events. We need a way of thinking, not about how one thing (drug) causes another (ideation), but how the drug in the context of an RCT amplifies certain kinds of ideo-affective organizations in some young people. I take the data on suicidal ideation to be not so much a warning of a latent, undocumented threat to the well-being of those undertaking biological treatments (Healy 2004; Liebert and Gavey 2009) as a sign of movement and reorganization within a system, strongly colored by affectivity, where the difference between a remedy and a harm is constantly in play.

The monoamine theory of depression suggests that low mood is caused by a depletion of serotonin in the CNS. SSRI antidepressants act as a remedy, it is presumed, by increasing the level of serotonin available for neurotransmission. There can, of course, be too much of a good thing. "Serotonin syndrome" is an adverse drug response involving a serotonergic agent. The typical clinical features of the syndrome are cognitive confusion, agitation, bodily twitching (myoclonus, hyperreflexia), shivering, tremor, sweating, fever, diarrhea, and/or poor coordination that have resulted from a toxic overload of serotonin in the central, enteric, and peripheral nervous systems (Sternbach 1991). The syndrome is usually the result of an overdose of a serotonergic pharmaceutical or an adverse reaction between a serotonergic agent and another drug. The syndrome has been closely linked to SSRIs (and the increased profile of the syndrome in the last decade no doubt reflects the significant increase in the prescription of serotonergic antidepressants), but it has also been associated with tricyclics, MAOIs, over-the-counter cough medicines, antibiotics, anticonvulsants, antimigraine medications, herbal products, and recreational drugs (Boyer and Shannon 2005). The syndrome has been identified in newborns, children, adults, and the elderly, and cases can vary from mild (tremor and diarrhea) to fatal (delirium, neuromuscular rigidity, hyperthermia) (Boyer and Shannon 2005).[9]

The problems caused by excess serotonin have been noted since at least the 1960s, when John Oates and Albert Sjoerdsma (1960) documented the toxic effects of tryptophan in patients receiving MAOI antidepressants. Most of the contemporary clinical literature is concerned with raising awareness of the symptoms among clinicians and debating the appropriate criteria for the diagnosis of the syndrome (Birmes et al. 2003; Boyer and Shannon 2005; Dunkley et al. 2003). For my purposes serotonin syndrome is noteworthy for two reasons: (1) the divergent symptoms of the syndrome—autonomic, muscular, integumentary, cognitive, and digestive—illustrate how some bodily functions, although remote, are nonetheless sympathetically attuned to each other through the actions of serotonin; (2) it muddles any clear notion we might have about whether serotonin is a poison or a cure.[10] Any monoamine treatment of depression needs to be sympathetic with, rather than impervious to, these disseminated, pharmakological conditions. The ways in which SSRIs fail to be entirely beneficial or entirely localized might be thought of as a measure of their efficacy: the failures and harms of SSRIs are just right—good enough—to find resonance with the system they inhabit. Imagining an antidepressant that had no adverse effects, we might begin to wonder if it would be effective at all. Could it modulate a depressive system that is made of adverse events, disappointments, and injuries? How could an entirely upright, self-contained drug engage the conditions of depressive despair?

An effective neuroscience of pediatric depressions needs to have a similar kind of pharmakological profile. Anita Miller (2007) offers a social neuroscience model of depression in children and adolescents that illustrates what such nonlinearity might look like when described in traditionally scientific terms. Reviewing data across many different studies, Miller draws our attention to an extensive, sympathetic network of depressive enaction in children and adolescents. First, we see that most major pediatric depressions (90 percent) are comorbid with another kind of psychological dysfunction (dysthymia, anxiety, bipolar), and the nature of these comorbidities varies with gender. Pediatric depression will also likely interfere with social, emotional, cognitive, and/or academic developmental milestones, and these deficits will feed back onto depressive states. Families with already existing mood disorders are evident in 94 percent of children with major depression, and genetic studies show increasing severity and earlier onset of depression across generations (genetic anticipation). From these data we can say that pediatric depression is an extensive, systemic event, contained neither to specific biomarkers nor to discrete symptomatic parameters nor to static demographic variables. In fact, the genetic research, so often thought of as the beating heart of biological reductionism, offers Miller a way to think about the dynamic, overdetermined character of depressogenic systems. She reports on genetic research on brain-derived neurotrophic factor (BDNF)—a protein involved in the regulation of neuron development and maturation, neuronal plasticity in adult brains, and stress resistance. Like serotonin, BDNF appears to be implicated in both the neurophysiology of depressive disorders and the pharmaceutical treatment of depression. In animal models, chronic stress leads to the down-regulation of BDNF in the hippocampus, which may lead to neuronal atrophy and reduced neurogenesis (Duman 2002).

Chronic use of antidepressants leads to an up-regulation of BDNF in the hippocampus and frontal cortex, perhaps intensifying the psychological effects of SSRIs and other antidepressants. That is, the one substance promotes the conditions for depression and the conditions for its treatment: BDNF is systemically available as both poison and cure.

These data demonstrate not a location ("static focal deficit"), but dynamism. They ought to alert us not to interaction between preexisting agencies (neuron × environment interactions), but to the way in which genes and environments are demarcated as agents together, out of the same intra-active fabric. Neurons and environment are siblings of coeval origin, not strangers who occasionally meet at scenes of pathology. Similarly, Miller argues that psychosocial context (for example, family discord) is less a static variable that determines pediatric depression than it is one mode of a dynamic developmental system through which early emotional dysregulation may culminate in depression in some high-risk children and adolescents. For Miller, studies of "isolated component processes often cannot capture this inherent feature of early-onset mood disorders" (51).

In particular, Miller takes systems of emotional regulation to be important sites of intensification for pediatric depressions. There is mounting evidence that one's style of emotional regulation is important to how behavioral, interpersonal, and cognitive capacities cohere into a depressive state: "Many depressed youngsters are self-focused and ruminative about their feelings and shortcomings. . . . They appear to perseverate in a prolonged pattern of negative emotions in that they tend to continue or repeat dysfunctional emotional behaviors even after the cessation of emotionally provocative events" (51). Bleak circumstances like relationship loss, witnessing family violence, and physical or sexual abuse are not in themselves directly depressive, but they may lead to prolonged affective stress that the adolescent and particularly the child will be unable to regulate and redress. Over time, it is this dynamic torpor (coupled with ongoing emotional strain) that can lead to serious disorders of mood. The neurobiological research on pediatric depression supports this hypothesis about the centrality of affective dysregulation. Miller discusses evidence from neuroendocrine, EEG, and neuro-imaging studies that suggests pediatric depressions emerge, in part, from dysfunction in neural systems that regulate emotion. In particular, it is the neurobiology of emotional processes that change as the child or adolescent develops that is most vulnerable during depressive episodes; these processes are intimately connected to other systems of sleep, food, and stress regulation. She concludes systemically: "Rather than correcting focal deficits, the key challenge may be finding ways to intervene in dynamic biological processes that are inherently shaped by the environment" (57).

It is the status of "inherent" in this conclusion that carries much of the conceptual weight in Miller's argument, and that is important for the case I am building here. If the "environment" (and in these passages that means any putatively non-neurological process: hormones, blood sugar, respiration, mother, siblings, school, socioeconomic status) is already (inherently) part of the dynamism of neurological processes, then we are no longer dealing with a theory that adds up various components (neurons + family + violence) as a way of calculating depressive outcomes. Miller asserts that her goal is "neither to reduce child and adolescent depression to isolated neurobiological measures nor to give broad generalizations with metaphorical terminology" (57). Might we be able to find some middle ground of conceptual advocacy in which the metaphoricity of depression and the neurobiology of depression cohabit, entwine, and are inherently shaped by one another? Miller offers a dynamic adaptive systems framework in which brain reward systems, gender, the biological changes of puberty, the development of the capacity for abstract thought, cortical regulation of emotion and attentional impairment together breed the conditions out of which pediatric depressions arise. Even if Miller has a tendency to occasionally long for "disentangling . . . causes and consequences" (47), and even if she remains largely focused on CNS activity rather than thinking about the minded character of the nervous system more broadly, she nonetheless constructs a schema of biological mutuality that is immensely helpful for those of us in the humanities and social sciences who, to paraphrase J. L. Austin (1962), want to know how to do things with pills.

Given the complexity of the neuro-depressogenic system that Miller sketches for us, can anything rightfully be excluded from it: pharmaceutical treatments, talk therapy, cultural heterogeneity, neurotransmission, metaphor? Fonagy's concern (with which I opened this chapter) to protect a mode of nonpharmaceutical psychological treatment seems isolationist in the light of neurology's natural capacity for xenoaffiliation, and out of step with Fonagy's own ambitions for more sympathetic theories of mind. My argument in this chapter has been that efforts to exclude or minimize the role of pharmaceuticals in the treatment of depression have the effect of taming the aggressive ambiguity of the pharmakon. This, in turn, feeds back onto treatment regimes, perhaps making them less directly harmful but at the same time also making them less capacious, less robust, and thus likely less effective.

Conclusion

It is, of course, especially hard to think of harm in relation to the treatment of pediatric populations. I am not thinking so much of the aversion to harming children, but of the much greater aversion to thinking of children themselves as harmful. Here Donald Winnicott (1949) is of considerable help. While he is widely known for this work on transitional objects (writing that tends to conjure images of well-worn blankets and teddy bears and satisfactorily nurtured children), Winnicott is also very well known, in clinical circles, for an essay on the curative effects of hatred for one's patients.[11] He begins "Hate in the Counter-Transference" by differentiating not between cure and harms, but between two different kind of harms that therapies may effect. On the one hand, Winnicott was an outspoken critic of electroshock therapies and psychosurgery for mental ill patients. While he could perhaps "forgive those who . . . do awful things" (69) because of the emotional toll that psychotic patients impose, he nonetheless sees such treatments as brutal, nontherapeutic harms that ought to be discontinued. On the other hand, there are injuries that the analyst cannot evade: "However much he loves his patients he cannot avoid hating them, and fearing them" (69). Winnicott argues that there are particularly troubled patients (like the ones described by Teicher, Glod, and Cole [1990]) that are captured by a state of "love-hate" (70), in which they fear that if the analyst is capable of love he is surely capable of murderous hatred. This demands of the therapist a direct relation to his own hatred: "Above all he must not deny hate that really exists in himself" (70).

One of Winnicott's primary examples in this paper is of a rather poisonous child who came under his care: "He was the most lovable and most maddening of children, often stark staring mad" (72). The boy, a war evacuee, was sent to Winnicott because of his truancy. Winnicott describes the time that this boy lived with him and his wife as "three months of hell" (72). Which is to say that Winnicott hated the boy. As in an analytic engagement with a psychotic patient, Winnicott needed to be able to develop the capacity to manage and express that loathing: "Each time, just as I put him outside the door, I told him something; I said that what had happened had made me hate him. This was easy because it was so true. I think these words were important from the point of view of his progress, but they were mainly important in enabling me to tolerate the situation without letting out, without losing my temper and every now and again murdering him" (73). Winnicott's refusal to sentimentalize the boy's interior world allowed his own hatred and the hatred of the boy to find their resonance with each other ("he needs hate to hate" [74]). These three months of misery find their efficacy not in the avoidance of injury but in its management. What Winnicott's paper leaves for us is a clinical idiom for the curative-poisonous character of talk-based therapies. The question for Winnicott is not whether or not you hate a patient (you do), but how any remedy that the dyad might be able to effect depends on how well such harms (phantasized or otherwise) are metabolized.

Might Winnicott's ideas about hostility provide a general orientation to the politics of pharmaceutical treatments? That SSRIs can further disrupt, amplify, or entrench suicidal ideation in children and adolescents is not evidence that these antidepressants are inherently poisonous to nonadult minds, that they are toxic intruders within an otherwise stable and naive system. Rather, an adverse effect is a sign that such minds are not benign to begin with: suicidal ideation is not just self-harm—it is also aggression against those closest to you (see chapter 3). The pediatric world is a harmed and harmful one, even in the best of circumstances. One of the key lessons from the successes and failures of pharmaceutical treatment of pediatric depression is this: first, cultivate your capacities in relation to harm.

So too for feminist theory. A feminist theory that tries to apprehend the harms that are native to its own conceptual and political actions is a more robust endeavor than one that tries, vainly, to make itself pure of heart. Such a theoretical stance takes up more room, it generates more possibilities (and thus more risks): it has more bite. My argument through these chapters is not that feminist theory should be able to recognize and neutralize its harms, or that such harms can be made valuable. Trying to keep the negativity of aggression intact, I have been arguing that politics is a broadly and bitterly constituted activity; it is not a synonym for amelioration. The key question, then, is not one of choosing between harm or remedy, or adjudicating on how much hostility or how much reform we are able to avoid or create. Rather, feminist theory has the more engaging task of developing ways to exploit and expand political terrains that are always pharmakological in character.

CONCLUSION

What survived was being open to or game for the encounter and all that
might be unbearable about it.

—Lauren Berlant and Lee Edelman, *Sex, or The Unbearable*

In January 2014 two researchers from the Alimentary Pharmabiotic
Centre at the University of Cork published a short article in the popular
science magazine *New Scientist* about "the mind-altering effects of gut
bacteria" (Cryan and Dinan 2014, 28). Certain kinds of gut bacteria, they
claim, may have a beneficial effect on mood. When mice were treated
chronically with the probiotic *Lactobacillus rhamnosus* they showed fewer
behavioral and biochemical signs of stress. The mechanism of this influence is unclear, but the researchers hypothesize that some bacteria—
psychobacteria, they call them—regulate the neurotransmitter GABA
via the vagus nerve (which connects the gut and the brain). GABA has
an inhibitory effect on the mammalian nervous system; by changing
how GABA is expressed in the brain these gut bacteria are thought to
have an anxiolytic (destressing) effect on the animals tested in these
studies. John Cryan and Timothy Dinan hedge their bets about the consequences of such research. On the one hand, these data raise the
possibility "of unlocking new ways to treat neurobehavioural disorders
such as depression and obsessive-compulsive disorder" (2014, 28). On
the other hand, "we are still a long way from the development of clinically proven psychobiotics and it remains to be seen whether they are
capable of acting like—or perhaps even replacing—antidepressants"
(2014, 29).

A lot of people emailed me this article, knowing that I was working with data about the gut and mood. The research described in the article is more oriented toward anxiety and GABA than depression and serotonin. Nonetheless, this probiotic research seems close enough to the concerns of *Gut Feminism*—but not entirely like them—that it can serve as a way to think about how my claims about biology, hostility, pharmaceuticals, and mindedness might unravel, tangle, and retwine beyond the final pages of this book. For example, it begins the work of thinking about how these arguments might fare when faced with data from other biological systems: not just serotonin but also GABA, not just neurology but also the immune and hormonal systems, not just the gut-brain connection but also the hypothalamic-pituitary-adrenal axis, not just humans but also bacteria and rodents (all of which are implicated in the original research [Bravo et al. 2011] on which the short, popular article is based). My expectation is that the readings I have been fashioning in the prior chapters can't be transported, in their entirety, into new empirical domains. These chapters don't provide templates; they exemplify rather than prescribe. The particularities of biological systems make a difference to the conceptual work that can be done with them: data matter, as do the methods by which they are generated.

To my mind, the ways in which certain empirical fields are not quite like other fields is not a problem to be solved but a conceptual opportunity to be intensified and engaged. In this respect *Gut Feminism* is resolutely anticonsilient—empirically and politically. It does not seek to unify, reconcile, or integrate the material at hand. The "linking of facts and fact-based theory across disciplines to create a common groundwork of explanation" (E. O. Wilson 1998, 8) is not what I have in mind. Instead I have used the readings in each chapter to generate a series of dissonant hypotheses that might be tested in other feminist venues: biological substrata have phantastic capacities; the biological periphery is implicated in minded states; feminism shuns its own hostilities; pharmaceutical treatments are a site of curiosity; cure cannot be detached from harm; negativity does not make good.

Before I say more about these claims and about the probiotic research in particular, let me say a few words about one of the conceptual environments into which this book arrives and which will likely influence the critical half-life of the arguments in these chapters. I am thinking of the so-called neurological turn in the humanities and social sciences. In recent years there has been increasing interest in how the humanities and social sciences might become more engaged with the natural sciences. The neurosciences have been perhaps the most heavily trafficked of the data sources: neuroesthetics, neuroeconomics, neurohistory, neurophilosophy, neuropsychoanalysis. In an excellent review of the neurological turn in the social sciences, Des Fitzgerald and Felicity Callard (2015) describe two commonplace ways in which neurological data have been addressed: on the one hand, there are those who launch a critique of "neurobiological chauvinism" (9) and defend the ongoing relevance (and perhaps the primacy) of sociocultural analysis; and on the other hand there are those who occupy a position of scientific ebullience, taking "experimental results and theoretical statements from the neurosciences as more-or-less true" (11). The neuro-skeptics and the neuro-enthusiasts: recto and verso. If *Gut Feminism* leans heavily on the weaknesses of this first position—that is, if it has been concerned that feminist antibiologism inhibits more than it expands our conceptual and political worlds—it is not therefore allied with the second of these stances. Indeed, over the life of this project, it has become my assessment that many of the new syntheses of the neurosciences and the humanities are too monochromatic in their ambitions (Wilson 2011). These new interdisciplinary ventures (including some feminist ones) strive for seamless knowledges and unified politics. As Fitzgerald and Callard note, the ambition of such research has been to recruit neuroscientific data in order to "*settle matters*" (20). Increasingly, these projects are generating an analytic monotony where neuroscientific data are used to cement what can count as the unconscious, or narrative, or economic behavior, or mental distress. They are generating something just as ruinous to the conceptual landscape as the Two Cultures debates: conceptual monocultures.

With Fitzgerald and Callard, and in the spirit of a different mode of interdisciplinarity, I hope that by bringing biology and feminist theory together each is a little undone in the encounter. My concern is not how biological inquiry could be diverted and rebuilt by feminist critique (although that remains an important project) but rather how feminist theory itself might be jostled and reanimated by an engagement with biology—particularly a phantastic biology and a biology of the periphery. Some kind of mutual rough-and-tumble strikes me as essential to any successful humanities-neurosciences alliance. Both the neuro-sciences and the humanities have to feel that something important is being destabilized. This alliance should feel vertiginous; it should not be mistaken for amity, solidarity, or amelioration. One of the key contributions that *Gut Feminism* has to offer to this kind of dissonant alliance is the concept of biological phantasy (the other contribution is an analysis of hostility, more of which below). Sándor Ferenczi's wild speculations about bodily materializations (a lump in the throat, a child in the stomach, a penis in the rectum) and amphimixis (anal-urethral-erotic admixtures) have been particularly important for encountering biology as a nonfoundational substrate. His speculations about a "third dimension" to biology (its capacity to be motivated and to think) and Klein's unruly account of the primordial-biological nature of phantasy have been crucial to reading biology as something other than dispassionate bedrock—without having to read against biology *in toto*. That is, a theory of phantasy has enabled me to attach to biology other than in a juridical register; it has given me the tools to start displacing an imperious, unyielding biology with one that (no less intricate and perhaps no less vicious) is keenly motivated, networked, and mobile. This biology-with-phantasy is a challenge to neuro-skeptics and neuro-enthusiasts of any disciplinary or interdisciplinary stripe. By remaining firmly engaged with biology, but moving against the conventions that configure biology (for better or for worse) as determinate matter, *Gut Feminism* works to breach the logics of both antibiologism and consilience.

How does the neurological turn structure responses to the research on bacteria and mood? In the first instance, there is plenty to irritate the neuro-skeptic. The research by Javier Bravo and colleagues (2011) is strongly oriented to the central nervous system, not just at the expense of social or cultural analysis but also at the expense of encountering the mind below the neck. Despite the bodywide distribution of GABA (Watanabe et al. 2002 document the evidence for GABA in the stomach and small intestines, kidneys, pancreas, pituitary, reproductive tissues, lungs, salivary glands, and optic nerve), Bravo and his colleagues do not think of anxiety as anything other than a cerebrally regulated process. This CNS chauvinism is evident in one of their most compelling experimental results. In order to test the hypothesis that L. *rhamnosus* influences mood by regulating GABA expression in the brain, Bravo et al. sever the vagus nerve in their mice (vagotomy). The vagus nerve is the main nervous conduit between gut and brain, and Bravo et al. argue that it is the route that allows enteric L. *rhamnosus* to regulate mood. The effects of vagotomy were clear-cut: it "prevented the anxiolytic effects of L. *rhamnosus* (JB-1) in mice" (16052). That is, the mice that lost vagal communication between the gut and the brain had significantly lower responses to the anxiolytic effects of the probiotics than those mice in which the vagus nerve was intact. For Bravo et al., these data confirm the idea that anxiety is regulated centrally (in the brain) rather than peripherally (in the gut). Despite their rhetoric of mind-body engagement, the skeptic might claim, Bravo et al. consolidate the primacy of the CNS over the peripheral body—effectively substituting a mind-body problem with a brain-body problem.

This conventional theory of mind is further exemplified, the neuro-skeptic could argue, in how Bravo et al. operationalize anxiety and depression in the mice—relying on the flat behavioral logic of the cognitive and pharmacological sciences. For example, one of the key behavioral measures for anxiety in their study is performance in the elevated plus maze (EPM). The EPM is a four-armed maze in the shape of a plus sign. The arms of the maze are long and narrow (10 cm wide, 50 cm long). Two of these arms are completely open (as if the mouse is walking the plank), the other two arms are enclosed by high walls (but no roof). The entire apparatus is elevated 50 cm from the floor. The EPM has long been established as a reliable method for calibrating anxiety and for measuring the anxiolytic effects of GABAergic drugs (e.g., benzodiazepines) in rodents (Pellow et al. 1985). The EPM is anxiogenic because, all things being equal, rodents will avoid open spaces and heights, exhibiting a preference for staying in close physical contact with environmental structures like walls. How often a mouse put all four paws into the open area of the EPM and how long it spent in that open space are taken by Bravo et al. to be measures of anxiety. Mice given L. *rhamnosus* entered the open arms of the maze more frequently, and stayed there longer, than mice without probiotic treatment: L. *rhamnosus* enables some mice to down-regulate anxiety (or is it fear?) sufficiently that they can explore a novel and aversive environment. In all likelihood, the neuro-skeptic (or the feminist antibiologist or the antipsychopharmaceutical advocate) will be critical of this experimental arrangement. She will find the theory of mind operationalized in the EPM insufficiently complex phenomenologically, and too focused on neurophysiology, to make comparisons to the culturally contingent and emotionally variegated nature of human anxiety credible. For the skeptic, this set of experiments looks like neuro-reductive business as usual.

The neuro-enthusiast is much more likely to take the results from this study without such qualifications, focusing instead on the possibilities these data offer for thinking nature and culture together. Fitzgerald and Callard (2015) note that such writers want "to assign to the natural or experimental sciences the task of generating findings that will confirm, verify and/or reveal the theoretical insights of cultural and social theory" (12–13). For example, Bravo and colleagues' data might point to the importance of dietary or homeopathic regulation of mental distress (instead of treatment by psychopharmaceuticals). Might these kinds of experiments bolster certain kinds of (body-oriented) alternative healing practices? Perhaps the microbiome-gut-brain axis is the beginning of a psychosomatic alliance between neurobiological researchers and cultural critics? Indeed, in the *New Scientist* account of this research Cryan and Dinan encourage this kind of reading: "At a time when prescriptions for antidepressants have reached record levels, effective natural alternatives with fewer side effects would be welcome" (29).

Such enthusiastic approaches to gut and brain are no less fraught than those of neuro-skepticism. One way in which neurological data are avidly used is to promote alternative perspectives on mind or mood; however, often this happens on the back of experimental methods so orthodox that even a cursory examination of the specifics would undercut their value for thinking mental distress anew. For example, can data extracted from the conventional parameters of the EPM (which equate maze-running behavior with emotional feeling, and which unhelpfully scramble the distinction between anxiety and fear) be used to think in novel ways about psychosomatic events? Doesn't the experimental focus on the linearity and exclusivity of vagal traffic make the rodent body too circumscribed (and too CNS focused) for a theory of psychosomatics? Just as frequently, ebullience about neuroscientific data has worked the other way round: the results of neurological research have been used in the humanities or social sciences in ways that reify conventional distinctions (between, say, drug and food, natural and artificial, remedy and harm) paying insufficient attention to how the data may, in fact, destabilize such partitions. For example, if GABAergic systems are not exclusively inhibitory but are also sometimes excitatory (Schuske, Beg, and Jorgensen 2004), how clearly are we able to draw a line between anxiogenic and anxiolytic events, or between the remedies and harms that emerge from the pharmaceutical regulation of GABA? In either case, I am arguing, the ebullient encounter between the neuroscientific and the critical misses its mark if it underreads empirical detail in order to generate a position of political or conceptual consilience.

Neuro-skepticism and neuro-enthusiasm are best thought of as stances that any argument (starting with this one) may occupy with varying degrees of intentionality or intensity. It has been the ambition of *Gut Feminism* to illustrate the considerable weaknesses of both such modes of reading and to find other pathways by which biological data can become critically mobile. However, the extent to which my analysis can extract itself from the lure of skepticism and enthusiasm is unclear. Any conceptual cut is also a bind. My distrust of both the neuro-skeptical and the neuro-ebullient positions cleaves me to them in ways that cannot be avoided. I will return to that critical predicament shortly. But for now, let me gesture briefly toward one approach in relation to GABA that tries to move athwart the conventional certainties of skeptical and ebullient positions. The ubiquity of GABA in the human body has led to investigations of how it might play a role in conditions as diverse as developmental disorders, sleep disorders, alcoholism, anxiety, schizophrenia, and Parkinson's disease (Olsen and Li 2012). Moreover, the extensive influence of GABA cannot be contained inside the human body; GABA is found in other animals—vertebrates and invertebrates—and also in plants. Indeed, Kim Schuske, Asim Beg, and Erik Jorgensen (2004)—studying the nervous system of the nematode *Caenorhabditis elegans* (which has been an exemplary site for GABAergic research)—find in GABA a way of demonstrating the intimacy of nervous action across vast periods of evolutionary time and between otherwise estranged biological systems. While worm and vertebrate lines diverged 800 million years ago, their nervous systems (and especially their GABAergic capacities) are still closely akin: "The components of the nervous system were not gradually perfected as more complex organisms arose. Rather, an elaborate nervous system evolved using the full complement of neurotransmitters. . . . This simple organism, perhaps a worm, then gave rise to more complex organisms, whose nervous systems differ little from those of the worm except in size" (413).

As Ferenczi would remind us, there is much that is wormlike in the human mind, and there are psychic functions (like anxiety) that fissure the organic world, disrespecting the divisions of species, genus, family, order, class, or phyla. Perhaps, then, there is a tendency to GABAergic "hyperexcitability" (Olsen and Li 2012, 367) that inheres in all organic substrata. Might the dispersal of GABA help us think about how there is not one location for anxiogenic and anxiolytic events (the gut, the brain, the social, the mother), but a dispersed nervousness that traffics (biologically) between gut and brain, and also across that crucial evolutionary divide of vertebrate and invertebrate, and among the organic kingdoms (animalia, plantae, bacteria). In particular, might the pervasiveness of GABA suggest an unbearable mentation in biology—a phantastic capacity for organic matter to become agitated or unglued? My point is not that neuroscientific research about GABAergic function *verifies* such capacities (and thus settles the matter of neurology and anxiety). Rather, I am arguing that close attention to some data about GABA might bring to light a systemic traffic in disquiet that cannot be fully grasped by either neuroscientific or critical texts, but can be glimpsed at these moments of disjuncture where neurological data and critical inquiry meet and cut across (cleave) each other.

Gut Feminism has not only been attentive to the nonconsilient nature of biology; it has also been keenly interested in how feminist theory can be written other than through the demand for political consilience and amelioration. I have been particularly focused on how feminist politics (like all politics) has an intrinsic hostility toward the objects, persons, and places it loves. While feminism is knowingly hostile to systems of injustice (sexism, homophobia, racism, wealth disparity), it is also hostile—in ways that cannot be extinguished—toward the things that it holds dear. This places important limits on how agreeable or enabling feminist politics can be; what might be most effective about feminism may have very little to do with its capacity to make good. In this regard, there is another important critical environment into which this book arrives: the recent debates about what has come to be called the reparative turn (Berlant and Edelman 2013; Hanson 2011; Love 2010; Wiegman 2014).

In an influential essay in 1997 Eve Kosofsky Sedgwick argues that the scene of queer, feminist, and poststructuralist criticism has become saturated with what she calls paranoid readings: "subversive and demystifying parody, suspicious archeologies of the present, the detection of hidden patterns of violence and their exposure" (21). Picking up the distinction between paranoid and depressive positions in the work of Melanie Klein, Sedgwick makes an argument for readings that draw on the repairing methods of the depressive position: "This is the position from which it is possible to use one's own resources to assemble or 'repair' the murderous part-objects into something like a whole. . . . [This] more satisfying object is available to be identified with and to offer one nourishment and comfort in turn" (8). A reparative reading would be more engaged with pleasure than suspicion, with amelioration than demystification. Heather Love (2010) notes that Sedgwick puts reparative reading on the side of "multiplicity, surprise, rich divergence, consolation, creativity and love," leaving paranoid reading to be the vendor of "rigid, grim, single-minded, self-defeating, circular, reductive, hypervigilant, scouringly thorough, contemptuous, sneering, risk-averse, cruel, monopolistic and terrible" analyses (237). Sedgwick's call to leaven paranoid readings with reparative ones has generated two widespread critical responses: first, to think that the splitting and hostility of paranoia is distinct from the ameliorative gesture of reparative readings; and second, to feel that reparative reading is an ethically more generous or gracious mode of analysis. This is where things begin to go astray. This is where *Gut Feminism* has tried to intervene.

That the conceptual field would become partitioned in this way (paranoid versus reparative) should not be a surprise to anyone, especially someone acquainted with the work of Melanie Klein. Sedgwick (1997) attempts to diminish the worst excesses of such partitioning: she repeatedly argues that paranoid and reparative readings are interrelated. However, while she argues that paranoid readings are infused by "powerful reparative practices" (8), she is more circumspect the other way round: the anxiolytic tendencies of reparative readings are "often" (8) in contact with the exigencies of paranoia, but not always. That is, she leaves open the possibility that there might be a reparative gesture that has no truck with the "hatred, envy and anxiety" (8) that structure paranoia. This idea of amelioration without hostility quickly became the ubiquitous understanding of what reparative reading means. In an interview with Sedgwick in 2000, for example, the interviewers refer to the "notion of a nonparanoid, reparative work" (Sedgwick, Barber, and Clark 2002, 258), seeming to presume that the reparative can operate in the absence of paranoid inclinations.

In a perceptive reading of Sedgwick's essay, Lee Edelman shows that Sedgwickian reparation cannot be so easily differentiated from paranoid hostility. Sedgwick's very first gesture of claiming a distinct conceptual and political space for reparation (consoling, not contemptuous) makes an analytic cut; and the oft-repeated figuration of oscillation between paranoid and reparative positions doesn't lessen the inevitability and the violence of this act. Instead, the legibility of reparation has been purchased through distancing it from the murderous splitting of objects that defines the paranoid position. By distinguishing between the paranoid tendency to divide and the reparative tendency to mend, Sedgwick's essay "repeats the schizoid practice it claims to depart from" (Berlant and Edelman 2013, 44). That is, her call for reparation makes its own "murderous division" (44): in splitting from paranoia, it enacts the self-defeating gesture of splitting from splitting. For Edelman what cannot be avoided, unbearable as it may be, is that reparation and amelioration are invested in some incisive cutting of their own. Indeed, what might be most hostile about reparation is its negation of its own hostility. Edelman's argument, as I understand it, is not that Sedgwick might have done better or done otherwise. He is not rebuking her for failing to have found a more scrupulously ameliorative reading practice. He is certainly not advocating for an oscillatory logic that would generate readings that are kinda paranoid, kinda reparative. Rather, he is illustrating the impossibility of finding a nonhostile space for our critical and political encounters. The issue is not that there should be no amelioration; it is that amelioration will always inflict some harm.

The desire for a clear distinction between the reparative and the paranoid (between remedy and harm, between skepticism and enthusiasm) has been particularly constraining for feminist engagements with psychopharmaceuticals. By believing in the separability of remedies and harms (and by making the pursuit of harmless remedies the goal of their interventions), such feminist readings have kept themselves at arm's length from the aggressive, bitter nature of depression and from a more thorough understanding of what any treatment and any theoretical stance entails. In this regard, they have been of a piece with a broader, debilitating feminist phantasy that our actions can turn to the good. In contrast, the feminism that *Gut Feminism* champions offers no plans for repair except through the interpretation of our ongoing, anxious implication in envies, hostilities, and harms. By this route I hope to have provoked some curiosity about what might happen in a political environment where conventional ambitions for amelioration or reparation lie gutted.

NOTES

1. Underbelly

1. I will be using the designation "phantasy" rather than "fantasy" in order to keep the specificity of British (Kleinian) psychoanalytic theory to the fore (rather than its French or American equivalent). Elizabeth Spillius et al. (2011) give a lucid account of these national/conceptual differences:

> Susan Isaacs . . . suggests the use of the "ph" spelling for unconscious phantasy and the "f" spelling for conscious fantasy. Some analysts have adopted Isaacs' suggestion, but most British analysts now use the "ph" spelling for both unconscious and conscious phantasies, at least in part because it is often difficult to be sure whether a patient's phantasy is unconscious, tacitly conscious or fully conscious. Laplanche and Pontalis criticise Isaacs' usage because in their view it disagrees with the profound kinship that Freud wished to emphasise between the conscious phantasy of perverts, the delusional fears of paranoid patients and the unconscious phantasy of hysterics. The spelling situation is further complicated by the fact that most American analysts use the "f" spelling for both conscious and unconscious phantasies. (5)

2. The feminist work on biology (or critical work on biology that has been informed by feminist theories of the body) is now extensive and diverse in terms of its political and conceptual ambitions: since 2000 see Alaimo and Hekman 2008; Alaimo 2010; Birke 2000; Bluhm, Jacobson, and Maibom 2012; Chen 2012; Cooper 2008; Fausto-Sterling 2000, 2012; Franklin 2007, 2013; Giffney and Hird 2008; Grosz 2004, 2005; Haraway 2007; Hekman 2010; Hird 2009; Jordan-Young 2010; Keller 2002, 2010; Kirby 2011; Mol 2002; Mortimer-Sandilands and Erickson 2010; Richardson 2013; Roberts 2007; Rosengarten 2009; Waldby and Mitchell 2006.

3. For excellent introductions to the work of Melanie Klein, see Hinshelwood, Robinson, and Zarate 1997; Likierman 2001; J. Rose 1993; Segal 1979; and Spillius et al. 2011.

4. Leo Bersani (1990) sees a closer theoretical affiliation between Klein and Freud than I am suggesting here: "Klein elaborated the most radical—at once the most compelling and most implausible—theory regarding infantile anxiety and aggression in the history of psychoanalysis. . . . [Her] scenarios of infantile violence, for all their apparent extravagance, rigorously and brilliantly spell out the consequences for our object relations of those destructive desires that Freud had already associated with infantile sexuality" (18–19). Jacqueline Rose (1993) recounts hearing commentary at an academic and clinical conference on Klein that Klein was not "a theorist in the strict sense of the term" (139). Rose reads Klein's theoretical mode slightly differently: "What happens if we read [that] comment not as a statement *against* theory, but as suggesting that Klein does theory *otherwise*, that Klein produced a theory, which, because of what it was trying to theorize, could not by definition contain or delimit itself?" (139).

5. For Jean Piaget (1929/2007), animism is one of the attributes of the preoperational stage (ages two to seven years). In this model animism is a cognitive style that a normally developing child will eventually outgrow.

6. For example, the question "Are SSRIs being overprescribed?" could become an investigation into how a pill is a technology of incorporation that meshes with (and calls forth) specific psycho-enteric responses to loss. What am I taking in when I swallow a pill? The widespread use of SSRIs then starts to look less like direct psychiatric malfeasance (domination) and more like a systemic (imbricated) attempt to regulate endemic melancholia (sometimes successfully, sometimes not). Notwithstanding the considerable pecuniary ambitions of Big Pharma, the pharmaceutical companies do not hold power over the treatment of depression: they are no more the origin of overprescription than their pills are the definitive cure of depressive symptomology. There are no bright dividing political lines that would allow us to know where to stand and what to stand for/against in relation to SSRI medications. This argument is laid out in detail in chapters 5 and 6.

7. Karl Abraham's work is an important bridge between classical Freudianism (orality and anality) and the stomach-animism of Kleinian phantasy. Abraham was Klein's second analyst and an important contributor to psychoanalytic theory: his early work (1911) was directly influential on Freud's "Mourning and Melancholia," and his later work (1924) is a clear precursor to Kleinian ideas about the sadistic nature of infant minds. In his extended essay on manic-depressive states, Abraham (1924) moves between a classically Freudian understanding of the symbolism of the body ("many neurotic persons react in an anal way to every loss, whether it is the death of a person or the loss of a material object. They will react with constipation or diarrhoea according as the loss is viewed by their unconscious mind" [426]), and a more phantastic understanding of the gut

("there exists [in the human embryo] an open connection between the intestinal canal (rectum) and the caudal part of the neural canal. . . . The path along which [psychical] stimuli may be transmitted from the intestinal canal to the nervous system might thus be said to be marked out organically" [499–500]). Klein was devastated by the sudden, early death of Abraham in 1925. She moved to London not long after, and her work there on phantasy, which she always claimed was deeply Freudian, seems to extend and intensify the lessons learned with Abraham in Berlin.

8. Susan Isaacs's paper on phantasy was the first presentation in the controversial discussions: it was discussed over the course of five meetings in the first half of 1943 and was published in 1948. Meira Likierman (2001) notes, "With this first presentation and debate, the Society discovered that the task of determining the psychoanalytic validity of Klein's ideas was unachievable. It was impossible to either collectively dismiss, or wholly accept Klein's thinking on unconscious phantasy, and the debate on it lingered without resolution for four months, before being abandoned for the second presentation and debate" (137). This paper became a classic in the field, and in the current literature Isaacs's position on phantasy is often taken to be interchangeable with Klein's (Spillius et al. 2011). Jacqueline Rose (1993) comments insightfully that Isaacs hovers in a "hybrid space of identification [with Klein] where bodies and psyches at once recognize each other as separate and get too close" (145). This relation (where, paradoxically, separation and being too close happen at the same time) speaks to what I would like to explore in relation to mindedness and physiology in Klein.

9. In the introductory lecture (23) titled "The Paths to the Formations of Symptoms," Freud notes:

> Whence comes the need for these phantasies and the material for them? There can be no doubt that their sources lie in the instincts; but it has still to be explained why the same phantasies with the same content are created on every occasion. I am prepared with an answer which I know will seem daring to you. I believe these *primal phantasies*, as I should like to call them, and no doubt a few others as well, are a phylogenetic endowment. In them the individual reaches beyond his own experience into primaeval experience at points when his own experience has been too rudimentary. (Freud 1917b, 370–371)

See Steiner 1991, 243 for an explanation of the uses of phylogenesis in Isaacs and Klein.

10. The *New Dictionary of Kleinian Thought* (Spillius et al. 2011), which updates Hinshelwood's canonical text, excises the commentary about the glittering moment in the history of the infant, but leaves Hinshelwood's conception of mind-body development intact: "The development of the human infant is a movement out of a world of bodily satisfaction into a world of symbols and symbolic satisfaction. There is a

progressive movement out of the body into the symbolic world" (406). Similarly, Likierman (2001) (another lucid and authoritative commentator on Klein) writes: "Klein offers a link between the blind, biological strivings of the young human organism, and the narrative, ideational faculties that emerge out of it" (139).

2. The Biological Unconscious

1. Hemianopsia is blindness in half the visual field (not blindness in the right or left eye, but blindness in the right or left side of each, or either, eye). It is a condition usually caused by a lesion to the optic nerves that carry information from each retina to the brain. The optic nerves partially cross over at the base of the brain (the optic chiasma) so that the nerves from the inner half of each retina (nearest the nose) cross to the opposite side of the brain. Consequently, objects in the right side of the visual field are projected to the left side of the brain; objects in the left side of the visual field are projected to the right side of the brain. Lesions to the optic nerves therefore cause blindness in only half the visual field. Hemianopsia can also be caused by cortical lesions; but Freud's argument is the same in either case: "What is in question in hysterical paralysis . . . is the everyday, popular conception of the organs and of the body in general. That conception is not founded on a deep knowledge of neuroanatomy but on our tactile and above all our visual perceptions. If it is what determines the characteristics of hysterical paralysis, the latter must naturally show itself ignorant and independent of any notion of the anatomy of the nervous system" (1893a, 170). Hysteria has no knowledge of the architecture of the optic nerves or the cortex, so it cannot produce a simulation of damage to these areas. Hysteria can only affect the everyday experience of vision (which is of an integrated, right-side-up visual field—producing total blindness in both or either eye).

2. Susan Bordo (1993), for example, in *Unbearable Weight*, cites Marx as the canonical figure who "reimagin[es] the body as a historical not merely a biological arena" (33). Gayle Rubin (1975), in her immensely influential paper "The Traffic in Women," used Freud and Marx and Lévi-Strauss to make the same kind of argument.

3. Perhaps the most distressing consequence of the breakdown in the Freud–Ferenczi relationship was the damage done to Ferenczi's reputation in the English-speaking world after his death (Bonomi 1998). In his biography of Freud, Ernest Jones claimed that while Ferenczi was the "most brilliant" member of Freud's inner circle, the one who was "closest to Freud," and a "gifted analyst" (Jones 1955, 178), he was nonetheless mentally unsound: "Toward the end of his life, [Ferenczi] developed psychotic manifestations that revealed themselves in, among other ways, a turning away from Freud and his doctrines. The seeds of a destructive psychosis, invisible for so long, at last germinated" (Jones 1957, 47). Jones misrepresented the relationship between Ferenczi and Freud at a time

when there was no public access either to Ferenczi's clinical diary or to the Freud–Ferenczi correspondence. He conflated Ferenczi's disappointment and anger at Freud with psychosis, and he took the psychological symptoms (such as paranoia) caused by neurological degeneration in the weeks before Ferenczi's death to be indicative of a latent, long-standing mental instability. Immediately following Ferenczi's death, Jones withdrew from publication (with Freud's approval) Ferenczi's famous last paper read at the 1932 Wiesbaden International Psychoanalytic Congress ("Confusion of Tongues between Adults and Child"), and he attempted to obstruct the later publication in English of Ferenczi's work (Balint 1988; Jones and Freud 2002, 720–722). In the wake of this damage to Ferenczi's reputation, the clinical diary was not published in English until 1988, and the letters were eventually published in English in 1992–2000 (Dupont 1988).

4. See Aron and Harris 1993; Haynal 2002; Rachman 1997; Rentoul 2010; Rudnytsky, Bókay, and Giampieri-Deutsch 2000; Stanton 1991; Szekacs-Weisz and Keve 2012; and special issues on Ferenczi in *Contemporary Psychoanalysis* 24 (1988); *Psychoanalytic Inquiry* 17.4 (1997); *International Forum of Psychoanalysis* 7.4 (1998); *American Journal of Psychoanalysis* 59.4 (1999); *Group* 23.3–4 (1999); *Journal of Analytical Psychology* 48.4 (2003); *American Imago* 66.4 (2009); and *Psychoanalytic Perspectives* 7.1 (2010). These days Ferenczi is variously cited as an important originator of ideas and techniques that led to object relations (Klein was first analyzed by Ferenczi), self psychology, and the contemporary interpersonal and relational psychoanalytic schools (via Clara Thompson and Michael Balint, both of whom were analysands of Ferenczi's). The New School for Social Research houses the Sándor Ferenczi Research Center. There is growing interest in Ferenczi in the nonclinical scholarly literature that draws on psychoanalytic theory (e.g., Thurschwell 1999).

5. In the preceding years (1916–1918) Freud and Ferenczi had discussed writing a book together about Lamarck: "The idea is to put Lamarck entirely on our ground and to show that his 'need,' which creates and transforms organs, is nothing but the power of Ucs. ideas over one's own body, of which we see the remnants in hysteria" (Abraham and Freud 2002, 361). This joint project never eventuated, but traces of this interest can be found in the later work of both men. Ferenczi's account of the regressive trend to earlier ontogenetic and phylogenetic states is explicitly Lamarckian in its sympathies (Ferenczi 1924, 50). Freud, eschewing all biological theory and data to the contrary, remained attached to the doctrine of acquired characteristics until the end of his life ("My position, no doubt, is made more difficult by the present attitude of biological science, which refuses to hear of the inheritance of acquired characteristics by succeeding generations. I must, however, in all modesty confess that nevertheless I cannot do without this factor in biological evolution" [Freud 1939, 100]). His Lamarckianism is most explicit in the posthumously published metapsychology paper "A Phylogenetic Fantasy." Freud's interest in Lamarck has been an

embarrassment for many orthodox commentators (Jones 1957; Sulloway 1979) but a point of interest to some sophisticated readers of psychoanalysis (Thurschwell 1999). Along with his interest in the occult (another shared enthusiasm with Ferenczi), Freud's Lamarckianism sits awkwardly with attempts to clarify what is (or should be) at the heart of classical Freudianism. On the other hand, Ferenczi's Lamarckianism seems more congruent with his intellectual ambitions and innovative clinical practice (Thurschwell 1999). Clearly, there was a divergence of intellectual preferences even in this shared project, and this may be one reason the work on Lamarck was never written.

6. Similarly, in 1923 Ferenczi writes to the *Internationale Zeitschrift für Psychoanalyse* repeating the following clinical anecdote from Bernheim's *Hypnotisme, Suggestion, Psychothérapie*:

> When I [Bernheim] was a pupil of M. Sédillot, that eminent master was called on to examine a patient who would not swallow any solid food. He felt in the upper part of the oesphagus, behind the thyroid cartilage, an obstruction at which level the alimentary bolus was retained, not regurgitated. On introducing his finger as deeply as possible across the pharynx, M. Sédillot felt a tumour which he described as a fibrous polypus projecting in the area of the oesophagus. Two distinguished surgeons touched it after him, and ascertained without hesitation the existence of a tumour such as the master described. Oesophagotomy was performed; no malformation existed at this level. (Ferenczi 1923, 105)

7. The OED gives the nineteenth-century meaning of *paraesthesia* as "disordered or perverted sensation; a hallucination of any of the senses." These days the term is reserved for an abnormal burning or prickling sensation that is generally felt in the hands, arms, legs, or feet but may occur in any part of the body.

8. Bulimia nervosa was isolated as a distinct disorder, separate from anorexia nervosa, in the DSM-III (1980). Bulimia is usually diagnosed when three criteria have been met: uncontrolled bingeing on large amounts of food; compensatory behaviors to rid the body of food (e.g., vomiting, laxative abuse, excessive exercise); and excessive concern about body shape and weight. The DSM-5 notes that many bulimic patients use compensatory measures (e.g., laxatives) to offset the weight gain from bingeing. Here I focus on vomiting (the most common of these compensatory methods), but there are a number of different ways to enact bulimia in the upper digestive tract (e.g., chewing and spitting out food rather than swallowing it; vomiting food back up into the mouth and then swallowing it again; see the discussion of rumination in chapter 3).

9. Binge episodes are usually measured through self-report, and most subjects are also given a battery of psychometric tests that measure depression and the severity of the bulimia. For a representative sample of this extensive literature, see Agras et al. 1987; Bacaltchuck and Hay 2003; Capasso, Petrella, and Milano 2009;

Fluoxetine Bulimia Nervosa Research Group 1992; Goldstein et al. 1995; Goldstein et al. 1999; Hughes et al. 1986; Leombruni et al. 2001; Pope et al. 1983; Shapiro et al. 2007; Zhu and Walsh 2002. Some of this research, especially in relation to the efficacy of fluoxetine hydrochloride (Prozac), was conducted by researchers employed or funded by Eli Lilly, the manufacturers of Prozac. Prozac is the only SSRI antidepressant approved by the FDA for the treatment of bulimia (although other SSRIs are prescribed off-label for its treatment).

3. Bitter Melancholy

1. The DSM notes that depression occurs more frequently in women than in men. Specifically, the DSM-5 observes that women have 1.5 to 3 times higher rates of major depressive disorder than men. The DSM-IV-TR noted that women are two to three times more likely to develop dysthymic disorder. The reasons offered for this disparity vary considerably. Perhaps it is hormonal imbalance: the DSM-IV-TR, for example, noted that "a significant proportion of women report a worsening of the symptoms of a Major Depressive Episode several days before the onset of menses" (APA 2000, 354). Perhaps it is maladaptive cognitive strategies: Nolen-Hoeksema (1990, 161) has argued that women are more prone to "rumination" ("cognitions and behaviors that repetitively focus the depressed person's attention on her symptoms and the possible causes and consequences of those symptoms"), and this makes it more likely that their depressive moods will be deepened and prolonged. Perhaps it is the effects of social injustice: "Biomedical and psychological theories of depression decontextualize what is often a social problem, simply acting to legitimize expert intervention, whilst negating the political, economic and discursive aspects of women's experience" (Ussher 2010, 15). Perhaps it is the effect of historical and discursive forces: "The historical disappearance of the category [male] melancholy has left only its devalued and quotidian counterpart, depression—which . . . is now viewed as a 'woman's complaint' " (Schiesari 1992, 95). To add to the complexity of the data, gender differences in depression are not seen in childhood; they emerge first in adolescence and are then stable through the adult and elderly years.

2. The "Freudian" hypothesis that depression is anger turned inward has gained widespread currency in psychological, feminist, and self-help literatures. See, for example, Frankel 1992, 39; Ingram 2009, 458; Nolen-Hoeksema 1990, 109; Salamon 2007, 151; Worell 2001, 145. There are differing opinions in the psychological literature about whether depressed patients are more or less hostile than nondepressed populations. I haven't pursued this literature here, as it relies on definitions of anger and hostility (conscious, reportable) that are quite different from psychoanalytic considerations of aggression (unconscious, libidinal). Moreover, this literature takes depression and aggression to be separable psychic events (depression may, or may not, be aggressive), whereas I am arguing that

aggression is an essential feature of depression; there is a hostility in melancholia that is foundational.

3. Here I use the terms *sadism, hatred, hostility,* and *aggression* interchangeably, and mean them to be (in part) unconscious and therefore irrational and libidinized. Elizabeth Spillius, Jane Milton, Penelope Garvey, Cyril Couve, and Deborah Steiner (2011) note that originally Klein defined *sadism* as the fusion of aggression and erotic phantasies. More recently, they observe, there is a tendency in the clinical literature to use sadism in a delibidinized sense (i.e., aggression). Anger, on the other hand, is an affect, so it has a different standing in relation to the primary processes. Silvan Tomkins (1991) argues that "an 'angry' response may or may not be an aggressive response. There is no *necessary* connection between anger and aggression, as a directed response. The infant may thrash about with flailing arms and limbs . . . [but] there is no evidence of any innate coordinated action intended to aggress upon the source of anger" (115).

4. There are many complications that need to be attended to when considering gender differences in depression. In a nuanced historical account, Laura Hirshbein (2009) argues that gender has been (unwittingly) woven into the contemporary psychiatric definition of depression that we are working with today: "Two important aspects of depression research in the 1970s and 1980s reinforced psychiatrists' conviction that depression was primarily a women's disease. First, researchers sought out ways to make their clinical trial samples of patients as homogeneous as possible and often focused exclusively on women patients in order to reduce variation in the patient population. Second, researchers eliminated features of the depression diagnosis that might have included more men" (99–100). Moreover, Steve Epstein (2007) argues, biomedical research on gender differences tends to reinforce conventional understandings of gender (and, I would add, conventional understandings of biology). As Anne Fausto-Sterling has shown, particular care is needed in relation to both method and conceptual frame when studying gender differences: "To develop dynamic approaches to embodiment, including the body's relationship to sex and to gendered social milieu, we need a starting point and a theory that can guide us as development proceeds. . . . Our framework integrates biology and culture in a fashion that has the potential to demonstrate the productive processes by which gender itself emerges and through which we can understand how seemingly sex-based differences in health are really due to the dynamic integration of biology and culture" (Fausto-Sterling, Coll, and Lamarre 2012a, 1684). Fausto-Sterling and her colleagues have been working with a developmental systems model to explain gender differences in behavior in young children. A schema of similar sophistication would be needed to interrogate gender differences in depression (which usually emerge only after puberty). To date, most research on gender differences in depression takes gender to be a self-evident axis of analysis, and this research usually has an orientation to nature-nurture interaction that consolidates orthodox notions of mind-body

(Oyama 2000). There are a number of different projects that could emerge out of a sustained, critical interest in gender differences in depression (see, for example, Radden 2003 and Schiesari 1992 for attentive historical analyses of gender and depression), but these lie outside the scope of the current project.

5. In an incisive reading of Sedgwick's essay, Edelman notes that reparation "repeats the schizoid practice it claims to depart from; but precisely by virtue of proclaiming its essential division from paranoia, and thereby erasing the division between paranoia's division and its own repair, it is able, ironically, to *enact* 'repair': the repair that undoes what we now can call *reparativity's* murderous division—and undoes it by way of its own implication in the negativity it seems to negate" (Berlant and Edelman 2013, 44).

6. Here Klein refers to psychic events for a very young child. She continues: "As the adaptation to the external world increases [as the child grows], this splitting is carried out on planes which gradually become increasingly nearer and nearer to reality. This goes on until love for the real and the internalized objects and trust in them are well established. Then ambivalence, which is partly a safeguard against one's own hate and against the hated and terrifying objects, will in normal development again diminish in varying degrees" (288). Given the coimplication of real and internalized objects and the dominating force of phantasy in the Kleinian system, it is clear that this approach to reality (and normal development) is asymptotic. Splitting, hatred, and ambivalence are diminished at later developmental points (all things being equal), but they are not eradicated. While children and adults usually become behaviorally more tame, they retain their phantastic viciousness indefinitely.

7. Some commentators make a historical or conceptual distinction between melancholia and depression (e.g., Radden 2003, 38–40; Salamon 2007, 151). Juliana Schiesari (1992), for example, argues that "melancholia and depression must be seen as differently constituted and legitimated in the *cultural field* according to gender. . . . Typically, when the loser is male, the loss *can* be idealized into the enabling conditions of his individualistic and otherwise inexplicable 'genius'; when the loser is female, loss becomes but a contingent circumstance in an essentialized and devalued depression. One task of the feminist analysis of melancholia is precisely to redeem the cause of depression, to give the depression of women the value and dignity traditionally bestowed on the melancholia of men" (93). Because the current argument is not primarily concerned with the questions of gender, history, and culture that occupy Schiesari's excellent analysis, I use the terms *depression* and *melancholia* interchangeably here and throughout.

8. This argument is even clearer in Klein's work, where much of the internal psychological landscape is made up of objects that have been introjected from the world and that retain their phantastic relation to external objects: "There is a constant interaction between anxieties relating to the 'external' mother . . . and those relating to the 'internal' mother, and the methods used by the ego

for dealing with these two sets of anxieties are closely interrelated. In the baby's mind, the 'internal' mother is bound up with the 'external' one, of whom she is a double" (M. Klein 1940/1975, 346). The distinction between self-hatred and hatred of others is much more convoluted in this scenario than it is in Freud, and the distinction between good and bad objects is complicated in profitable ways. In both cases, however, there is clear traffic of hostility between self and (internal/external) others. Both Klein and Freud were influenced in their thinking on melancholia by Karl Abraham (1911), who clearly understood that depressive hatred was directed out to the world: "The tendency such a [melancholic] person has to adopt a hostile attitude towards the external world is so great that his capacity for love is reduced to a minimum. At the same time he is weakened and deprived of his energy through the repression of his hatred or, to be more correct, through repression of the originally overly-strong sadistic component of his libido" (139).

9. See J. Rose 1993 for an analysis of the importance of this aspect of Klein's work, especially as it played out during the controversial discussions.

10. There are a number of compelling uses of Klein's work in the feminist and critical literatures. See, for example, Eng and Han 2006 on racial reparation; Giffney 2008 on the death drive and the human; Sánchez-Pardo 2003 for a comprehensive reading of Klein on aggression, gender, and sexuality. Lyndsey Stonebridge and John Phillips's (1998) anthology contains essays by Leo Bersani, Jacqueline Rose, and Judith Butler that have been important for the arguments I make here. Nonetheless, most of the contemporary feminist uses of Klein are interested in the relevance of her work for theories of gender, sexuality, and subjectivity. I am less interested in questions of identity or subjectivity than in the bodily systems that are constitutive of mind, and how Klein is particularly galvanizing for theorizing biology in terms that are phantastic and hostile.

11. The pleasurable character of merycism in infants is also noted by Clouse et al. 1999 and Geffen 1966.

12. Take, for example, Jane Gallop's provocative opening to Thinking through the Body, which was widely influential on the first wave of feminist theories of the body. She opens with an extract from Adrienne Rich's Of Woman Born: " 'On June 11, 1974, "the first hot day of the summer," Joanne Michulski, thirty-eight, the mother of eight children . . . took a butcher knife, decapitated and chopped up the bodies of her two youngest on the neatly kept lawn of the suburban home where the family lived outside Chicago.' . . . A mother brutally murders her babies, the youngest two months old. The act seems inhuman; the perpetrator a monster. Yet Rich tells us that, in 1975, 'every woman in that room who had children . . . could identify with her' " (Gallop 1988, 1–2).

13. Julia Kristeva (1989) lists merycism as one of several manifestations of grave psychosomatic disintegration in infancy: it prompts one "to accept the idea of a death drive that, appearing as a biological and logical inability to transmit

psychic energies and inscriptions, would destroy movements and bonds" (17). This gesture toward merycism is in the context of a very interesting distinction that Kristeva makes between two different kinds of melancholia: on the one hand, classical Freudian melancholia that arises from a loss/death and that has a hostile cannibalistic nature; on the other hand, a melancholia colored by narcissism in which the individual is troubled not by the loss of an object but rather by a profound, empty sadness: "Persons thus affected do not consider themselves wronged but afflicted with a fundamental flaw, a congenital deficiency. Their sorrow doesn't conceal the guilt or sin felt because of having secretly plotted revenge on the ambivalent object. Their sadness would be rather the most archaic expression of an unsymbolizable, unnamable narcissistic wound" (12). Kristeva's interests bend toward primal masochism rather than sadism or outwardly directed hostility, and she is most captured by ideational or linguistic events; for these reasons I don't engage with her work here. Moreover, her use of neurological data leans quite conventionally on a separation between the biological and ideational spheres: "And yet there is nothing today that allows one to set up any relation whatsoever—aside from a leap—between the biological substratum and the level of *representations*" (39). The difficulties with such a stance (a leap) have been articulated at length in these first three chapters, and I won't repeat those arguments in relation to the specifics of her work. A more sustained engagement with Kristeva's work could begin with the idea that the notion of a narcissistic mode of melancholia might not point to a different form of the illness, but to a different register of melancholic action (i.e., organ speech).

14. See, for example, Freud's (1931) account of the little girl's pre-oedipal, possibly gender-forming, hatred of her mother.

15. In 2011, McInerney pleaded guilty to second-degree and voluntary manslaughter and was sentenced to twenty-one years in prison.

4. Chemical Transference

1. In this chapter I keep my focus on the SSRIs and leave to one side analysis of the earlier-generation antidepressants (e.g., the MAOIs and the tricyclics) and the newer, atypical antidepressants that inhibit the reuptake of norepinephrine and/or dopamine as well as serotonin (e.g., bupropion/Wellbutrin, venlafaxine/Effexor). I maintain this narrower focus for two reasons. First, much of the critical and feminist literature that I am responding to has emerged in direct response to the SSRIs (particularly fluoxetine/Prozac). Second, as will become evident shortly, there is so much pharmacokinetic variation within the class of SSRIs that it becomes difficult to manage more data from outside that group. Crucially, I want to keep my analysis closely tied to the specifics of SSRI action and leave open the possibility that the differences in pharmacological action in different drug classes might elicit different kinds of analytic stances.

2. Here Ann Cvetkovich's (2012) work on depression has been exemplary. In an early account as she was beginning the depression project she wrote, "I don't believe in Prozac. No, I think it's a scam, even if that makes me one of those quacks, like people who don't believe that the HIV virus causes AIDS. Discussion about the biochemical causes of depression might be plausible, but I find them trivial. I want to know what environmental, social, and familial factors trigger those biological responses—that's where things get interesting" (15). As the project developed, she seemed to soften the tone of her manifesto: "I'm not against pharmaceuticals for those who find they work. . . . I do, though, want to complicate biology as the endpoint for both explanations and solutions, causes and effects" (16). What is important for the argument of this chapter is that attempts to "complicate biology" have just about always meant turning away from it and toward environmental, social, and familial data. Against this inclination, I want to show evidence of the complicated character of biochemistry. For me, that's where things get interesting. In this vein, see the excellent work of Mariam Fraser (2001, 2003, 2009) on serotonin.

3. Fluoxetine/Prozac is also manufactured in liquid form. The SSRIs and the other atypical antidepressants are manufactured only in oral form (Potter and Hollister 2001). Some of the well-established tricyclic antidepressants (e.g., imipramine/Tofranil, amitriptyline/Elavil) can be administered by injection: "Intramuscular administration of some tricyclic antidepressants (notably amitriptyline and clomipramine [Anafranil]) can be performed under special circumstances, particularly with severely depressed, anorexic patients who may refuse oral medication or ECT" (Baldessarini 2001, 463).

4. The bioavailability of an orally administered drug (i.e., its concentration in blood plasma) is measured as a percentage of the bioavailability of the same drug if it had been administered intravenously (IV). By definition, an IV-administered drug has a bioavailability of 100 percent, as it has avoided both the gut and the liver and is fully available in the blood. Usually a group of subjects are given single intravenous and oral doses of the drug on separate occasions: the bioavailability of the oral formulation is simply calculated as a proportion of the amount available when administered intravenously.

5. For critics like Peter and Ginger Breggin (1994) this distribution of antidepressant effect beyond narrowly defined serotonergic pathways in the brain is one of the signs that drugs like Prozac are toxic substances. They see the dissemination of drug effects as a kind of tyranny:

> Prozac has been shown to interfere with the functions of serotonin throughout the body, including the platelets in the blood, accounting in part for its wide variety of side effects.
>
> Overall, Eli Lilly's promotional line about Prozac's selective effects on the nervous system should be viewed with caution and skepticism. No one

prescribing or receiving the drug can fully grasp Prozac's overall impact on the brain and whole body, because it's beyond our current scientific understanding. (26)

This kind of claim is politically ineffectual, in my view, because it doesn't use this dissemination of drug effect to rethink conventional models of biological substrate and psychological cause and effect. After all, the allegedly toxic effects of antidepressants are no less illuminating than their supposedly therapeutic effects in terms of providing tools for thinking about mind-body entanglement. To the extent that the Breggins accept a very conventional model of direct and unwavering lines of influence from drug to behavior, they are more faithful to mainstream biopsychiatry than they suspect.

6. María Hernández and Appu Rathinavelu (2006, 198) write:

> The binding of a drug to a receptor molecule is analogous to a key (the drug) fitting into a lock (the receptor's binding site), and the ability for the ligand to activate a receptor can be compared to the ability of a key to open a compatible lock. . . . Therefore, to fit into the binding site, the ligand (drug) must also have a matching or compatible structure and the suitable three-dimensional configuration necessary for the binding. When these requirements are met, the ligand fits into the receptor's binding site like a key fitting a lock. Hence, this model of interaction is known as lock and key model.

SSRIs fit, like keys, into transporters that facilitate the reuptake of serotonin in the synapse, but they don't "open" the lock; rather, they block the ability of that transporter to remove serotonin from the synapse. In contrast, the serotonin receptors ("locks") in the pre- and postsynaptic membranes are directly triggered ("unlocked") by a drug like lysergic acid diethylamide LSD (Szabo, Gould, and Manji 2009).

7. Emmons's association of whole-body approaches with "the more naturalistic notions of balance common to humoral or homeopathic paradigms" (109) draws too clear (and conventional) a line between the natural and the unnatural, between the human and the inhuman (mechanical or chemical). By dividing the human from the mechanical in this way, she contributes to (and strengthens) the reductive metaphors of mechanism that she is trying to negate. Similarly, her gesture toward naturalistic balance underestimates the importance of imbalances (differentials, asynchronies, asymmetries, frictions) for psychological well-being (see chapter 3). It is my contention that psychological stability emerges from differentials across cell membranes, between (say) the conscious and the unconscious, and among cognitive strategies and affective surges and hormonal recalibrations. In this sense, dynamism (imbalance) is to be found at every level of the body.

8. See chapter 3 for an argument about how this turn from the world in melancholia isn't absolute (i.e., some libido is also turned, as aggression, onto

external objects). It was this difference between internal and external cathexis of libido that was initially thought to render certain narcissistic/psychotic disorders untreatable psychoanalytically (unable to cathect the external world, the melancholic would be incapable of engaging the clinician long enough to foster a transferential relationship). That clinical distinction about treatment no longer pertains (Laplanche and Pontalis 1988).

9. He is also one of the most influential. At the time of writing (July 2014) Ogden's paper on the "analytic third" is the second most cited article (in the last twenty years) in the Psychoanalytic Electronic Publishing database (the extraordinarily comprehensive database of psychoanalytic writing in English that contains all of the *Standard Edition of the Complete Psychological Works of Sigmund Freud* and the complete back catalogs of over fifty psychoanalytic journals). This ranking puts Ogden's paper in the company of Bion's paper on attacks on linking, Winnicott's on transitional objects, and Klein's on schizoid mechanisms.

10. It is no doubt worth reiterating here that the subjectivities of the analyst and analysand are also not singular or delimitable: they are internally cleaved by the unconscious, and they introject the external world. In the analytic session these subjectivities are also asymmetrically aligned: Ogden doesn't advocate disclosure of countertransferential feelings and thoughts.

11. For example, after the patient turned with "a look of panic on her face" (14) in response to Ogden's movement, he writes:

> It was only in the intensity of this moment, in which there was a feeling of terror that something catastrophic was happening to me, that I was able to name for myself the terror that I had been carrying for some time. I became aware that the anxiety I had been feeling and the (predominantly unconscious and primitively symbolized) dread of the meetings with Mrs. B (as reflected in my procrastinating behavior) had been directly connected with an unconscious sensation/fantasy that my somatic symptoms of malaise, nausea and vertigo were caused by Mrs. B, and that she was killing me. (14–15)

The question of destructiveness and aggression is mentioned only in passing in this chapter. I have dealt with these questions at length in chapter 3. Here I simply wish to note that the permeable relations I am describing here are not always benign.

12. Peter Kramer's *Listening to Prozac* (1993) was an early analysis of the changes in the psychocultural landscape that followed the release of Prozac into the US market in 1988. While widely quoted, this text (to my mind) has also been generally misunderstood as advancing a reductively biologistic view of depression. *Listening to Prozac* offers a nuanced account of how a widely used drug like Prozac alters the conceptual, clinical, and political demands that we are used to living with: "The change is not a matter of 'taking biology into account,' as if one can

maintain old ideas about behavior and personality and tack on a separate biological point of view. Medication has a pervasive influence, changing the way we see people and understand their predicaments" (285). I discuss the Lucy case history in more detail in *Psychosomatic: Feminism and the Neurological Body* (Wilson 2004).

13. The *Oxford English Dictionary* states that *sympathy* derives from the Greek "to have fellow feeling," and it gives as its first definitions "a (real or supposed) affinity between certain things, by virtue of which they are similarly or correspondingly affected by the same influence, affect or influence one another (esp. in some occult way), or attract or tend towards each other," and, in relation to pathology, a "relation between two bodily organs or parts (or between two persons) such that disorder, or any condition, of the one induces a corresponding condition in the other."

14. Angell notes elsewhere that one of the authors of this study reportedly made $500,000 in one year consulting to pharmaceutical companies that make antidepressants (Angell 2005). In addition, the demographics of the patient group, while perhaps typical of populations for psychiatric research on depression, are not reflective of the general population: the patients were predominantly female (65 percent) and white (90 percent), and most had extensive histories of treatment for depression (only 10 percent had never been treated for depression). The study also placed extensive restrictions on the kind of symptomology that could be tolerated in the study (again, this is typical of clinical psychiatric research): patients were excluded if they had comorbidities with seizure, schizophrenia, obsessive-compulsive disorder, bipolar disorder, eating disorders, borderline personality disorder, posttraumatic stress disorder, social phobia, or anxiety disorder, if they had tested positive for drugs of abuse, or if they had not responded in a previous trial to either Serzone or psychotherapy. The patients could not be taking anxiolytics or sedatives for sleep, and the female patients of childbearing potential had to agree to take contraceptives to prevent pregnancy during the study.

15. This general finding of the efficacy of "combined treatment" is of course somewhat heterogeneous. For example, the type of therapy that is used in the trial will affect the strength of the response rate, as will the severity of depression in the patient population. Even so, the National Institute for Health and Clinical Excellence (UK), the American Psychological Association, the American Psychiatric Association, the International College of Neuropsychopharmacology, and the British Association for Psychopharmacology all recommend combined therapies for the treatment of moderate to severe depression (Anderson, Nutt, and Deakin 2000; Sartorius et al. 2007).

16. Of course, a good working alliance doesn't always traffic in good feeling. Indeed, it is probably the capacity for the dyad to recognize and metabolize bad feeling and disjunctions in attunement that help promote successful therapeutic outcomes. And what constitutes a successful outcome is highly contingent on the circumstances of the treatment.

17. New capacities for permeability may also make things worse: permeability may mutate, denature, or corrode. In the final chapter I will explore how a cure (*pharmakon*) is never simply beneficial: it also involves modalities of harm.

5. The Bastard Placebo

1. They make this explicitly a matter of primitive versus modern technique: "A reasonable assumption from the study of primitive societies and early medical records is that placebos were the dominant treatment in preliterate cultures" (3); "medical reasoning about treatment languished during the seventeenth century in a primitive state: the lungs of a fox, a long-winded animal, were given to consumptives; the fat of a bear, a hirsute animal, was prescribed for baldness" (22); "treatment in primitive cultures did not differentiate among physical, emotional, and spiritual illnesses" (53).

2. *Kettle logic* is a phrase used to describe a kind of psychic defense first outlined by Freud in *The Interpretation of Dreams*. In his analysis of his own dream (the now famous dream of Irma's injection), Freud says: "The whole plea—for the dream was nothing else—reminded one vividly of the defense put forward by the man who was charged by one of his neighbours with having given him back a borrowed kettle in a damaged condition. The defendant asserted first, that he had given it back undamaged; secondly, that the kettle had a hole in it when he borrowed it; and thirdly, that he had never borrowed a kettle from his neighbour at all" (Freud 1900, 119–120).

3. "Placebo nation," *New York Times*, March 21, 1999; "No prescription for happiness: Could it be that antidepressants do little more than placebos," *Boston Globe*, October 17, 1999; "Against depression, a sugar pill is hard to beat: Placebos improve mood, change brain chemistry in majority of trials of antidepressants," *Washington Post*, May 7, 2002; "Make-believe medicine. Do placebos work? New research suggests they work surprisingly well—in fact rather better than some conventional drugs," *Guardian*, June 20, 2002; "Anti-depressants have little more effect than placebos, claims study," *Sydney Morning Herald*, October 21, 2002.

4. The Hamilton Depression Rating Scale (HAM-D) has become the most commonly used method for psychological assessment of depression in clinical trials (Hamilton 1960). It comes in different versions: a sixteen-, seventeen-, nineteen-, twenty-one-, or twenty-four-item checklist. The HAM-D is filled in by the clinician; it documents the severity of somatic and psychological symptoms such as insomnia, gastrointestinal disruption, agitation, mood, suicidal ideation, guilt, and anxiety. The criteria from the UK regulatory authority NICE (National Institute for Clinical Excellence) for clinical improvement are a shift of three points on this scale (Khan et al. 2010).

5. The 2008 study uses the same data set as in 1998 and 2002 but looks specifically at how different kinds of depression elicit different kinds of placebo

response. The study concludes that the difference between drug response and placebo response increases depending on the severity of the depression. That is, there is almost no difference (in the RCT data) between drug response and placebo response in cases of mild depression. There is a relatively small difference between drug response and placebo response in severely depressed patients. While these differences might be statistically significant in any given trial, they do not meet the criteria for clinical significance (three points on the HAM-D). The difference between drug and placebo response only becomes clinically significant in the trials that studied "the most extremely depressed patients" (1280). Moreover, even in these cases, Kirsch et al. argue, the difference is due not to greater drug response but to deflated response to placebo in this population (Kirsch et al. 2008).

6. These writers argue that the pharmaceutical companies generate diseases (e.g., social anxiety disorder) out of ordinary personality characteristics (shyness) in order to widen the market for their products:

> For example, social anxiety disorder—the fear of being embarrassed or humiliated in public—was considered a rare disorder until physicians began treating it with Nardil (phenelzine) in the mid-1980s and then, later, with SSRIs such as Paxil. Today social phobia is often described as the third most common mental disorder in the United States. Similar stories can be told for obsessive-compulsive disorder and panic disorder (the latter was known among clinicians in the mid-1980s as the "Upjohn illness," after the makers of Xanax). As David Healy has pointed out, the key to selling psychoactive drugs is to sell mental disorders. (Elliott 2004, 5)

7. See the website at http://www.placebo.ucla.edu.

8. A major depressive episode is defined in the DSM-IV (the guidelines under which Leuchter's research was conducted) as including five or more of the following within a two-week period: depressed mood; loss of interest in all or most activities; significant weight loss or gain; insomnia or hypersomnia; psychomotor retardation or agitation; fatigue; feelings of worthlessness or guilt; inability to concentrate or think; and suicidal ideation.

9. Florian Holsboer (2008, 641) writes:

> The many interactions within and between genes, as well as insertions and deletions, copy-number variations and, most of all, variations in non-coding DNA sequences, follow regulatory principles that are far from understood. This notion is strengthened by the fact that inherited variation in both gene and non-gene regions of DNA is strongly affected by epigenetic modifications that are inherited or that occur after conception. Thus, translation of a single genetic variation into a drug-discovery programme might not be a feasible gateway to personalized medicines. This is particularly true for

antidepressants, because single genes that have big effects, such as the BRCA gene variations that raise breast cancer risk by up to 85%, are unlikely to have a role in depression.

6. The Pharmakology of Depression

1. *Contraindication* means slightly different things in the United Kingdom and the United States. In the United Kingdom (the home of the Whittington article), contraindication means that a drug can be prescribed to pediatric patients only by a subspecialist (implying that its use should be minimized and carefully regulated). In the United States, contraindication means that the drug should not be used at all by pediatric patients (Leslie et al. 2005).

2. Karen Barad defines *intra-action* in contradistinction to the more familiar term *interaction*. Intra-action is "the mutual constitution of entangled agencies. That is, in contrast to the usual 'interaction,' which assumes that there are separate individual agencies that precede their interaction, the notion of intra-action recognizes that distinct agencies do not precede, but rather emerge through, their intra-action. It is important to note that the 'distinct' agencies are only distinct in a relational, not an absolute, sense, that is, agencies are only distinct in relation to their mutual entanglement; they do not exist as individual elements" (Barad 2007, 33).

3. Black box warnings are printed on the paper insert to be found inside the packaging of prescription drugs sold in the United States. The name *black box* refers to the black border around the text that issues the warning about possible serious adverse responses to the drug. Black box warnings are the most serious level of notification that the FDA can issue (short of withdrawing a drug from the market).

4. The 2006 study by Tarek Hammad, Thomas Laughren, and Judith Racoosin examined data from "4582 patients [in] 24 pediatric trials of 9 antidepressant drugs. Most of the trials were conducted in the 1990s, and trial durations ranged from 4 to 16 weeks" (335). From this pool they found 120 suicide-related adverse events (e.g., suicide attempts, suicidal ideation, preparations for suicide), but no completed suicides.

5. The studies that track pediatric use of SSRI antidepressants consistently show increasing rates of prescription of antidepressants to children and adolescents in the 1990s (Delate et al. 2004; Zito et al. 2003). There are, however, significant variations within this trend: before puberty boys and girls are prescribed antidepressants in roughly equal numbers, but after puberty girls are prescribed antidepressants much more frequently than boys (a disparity that continues on in adult populations); a sizable minority of pediatric patients appear to be prescribed the drug only once; a lot of antidepressant prescriptions are being used

off-label (e.g., for anxiety or nocturnal enuresis); about half of pediatric patients stop taking antidepressants after two months (Murray, de Vries, and Wong 2004).

6. *Fantastic Voyage* (directed by Richard Fleischer, 1966) narrates the journey of a manned nuclear submarine that has been shrunk to microscopic size in order to be injected into the bloodstream of a patient. The crew of the submarine have one hour to get to the patient's brain and destroy a blood clot. As one might expect, obstacles are encountered, harms are inflicted, but in the end a remedy is found and the patient is saved.

7. For a compelling ethnography that attempts to portray precisely this richness, in adolescents undergoing pharmaceutical treatment (buprenorphine) for drug dependency, see Meyers 2013.

8. For example, it is worth noting that while there is evidence that antidepressants raise the risks of suicidality in children and adolescents in RCTS, there is no evidence in autopsy studies that antidepressants are associated with completed suicides in children and adolescents. Moreover, since the early 1990s there has been a significant (31 percent) decrease in suicides in adolescent boys (Hammad, Laughren, and Racoosin 2006).

9. These particular sedimentations are not confined to humans: serotonin (5-hydroxytryptamine) is a neurotransmitter found in a wide variety of animals (for example, insect venom) and in plants (seeds and fruit). Even within the human body, serotonin is thought to be involved in the materialization of many different events: mood, appetite, sleep, pain, migraine, and vomiting.

10. In a chapter that challenges the "broken-brain" model of depression, Lennard Davis (2013) notes in a rhetorical aside that "in the case of serotonin, apparently you can't have too much" (49). Davis's parenthetical comment is a great example of where attention to empirical detail (in this case, interest in what happens when there is too much serotonin) changes the kinds of political and conceptual arguments that can be made. My claim throughout this book has been that some care for, and engagement with, the empirical literature would transform much of the now routine (and ineffective) antipsychiatric rhetoric in these kinds of analyses of SSRI antidepressants.

11. Winnicott's argument about hatred is in relation to psychotic patients (whom he differentiates from neurotic ones). The category "psychotic" was more capacious in 1949 than it is today and technically included melancholic conditions (see chapter 3).

REFERENCES

Abraham, Karl. 1911. Notes on the psycho-analytical investigation and treatment of manic-depressive insanity and allied conditions. In *Selected papers of Karl Abraham*, translated by Douglas Bryan and Alix Strachey, 137–156. London: Hogarth.

Abraham, Karl. 1924. A short study of the development of the libido, viewed in the light of mental disorders. In *Selected papers of Karl Abraham*, translated by Douglas Bryan and Alix Strachey, 418–501. London: Hogarth.

Abraham, Karl, and Sigmund Freud. 2002. *The complete correspondence of Sigmund Freud and Karl Abraham, 1907–1925*. Edited by Ernst Falzeder; translated by Caroline Schwarzacher. London: Karnac.

Agras, Stewart, Barbara Dorian, Betty Kirkley, Bruce Arnow, and John Bachman. 1987. Imipramine in the treatment of bulimia: A double-blind controlled study. *International Journal of Eating Disorders* 6.1: 29–38.

Alaimo, Stacy. 2010. *Bodily natures: Science, environment, and the material self*. Bloomington: Indiana University Press.

Alaimo, Stacy, and Susan Hekman. 2008. *Material feminisms*. Bloomington: Indiana University Press.

Anderson, Ian M., David J. Nutt, and John F. W. Deakin. 2000. Evidence-based guidelines for treating depressive disorders with antidepressants: A revision of the 1993 British Association for Psychopharmacology guidelines. *Journal of Psychopharmacology* 14.1: 3–20.

Angell, Marcia. 2000. Is academic medicine for sale? *New England Journal of Medicine*, 342.20: 1516–1518.

Angell, Marcia. 2005. *The truth about the drug companies: How they deceive us and what to do about it*. New York: Random House.

Antonuccio, David, David Burns, and William Danton. 2002. Antidepressants: A triumph of marketing over science? *Prevention and Treatment* 5.1: n.p. Available at http://psycnet.apa.org/journals/pre/5/1/.

APA (American Psychiatric Association). 1952. *Diagnostic and statistical manual of mental disorders* (DSM-I). Washington, DC: American Psychiatric Association.

APA (American Psychiatric Association). 1968. *Diagnostic and statistical manual of mental disorders*, 2nd ed. (DSM-II). Washington, DC: American Psychiatric Association.

APA (American Psychiatric Association). 2000. *Diagnostic and statistical manual of mental disorders*, 4th ed., text revision (DSM-IV-TR). Washington, DC: American Psychiatric Association.

APA (American Psychiatric Association). 2013. *Diagnostic and statistical manual of mental disorders*, 5th ed. (DSM-5). Arlington, VA: American Psychiatric Association.

Aron, Lewis. 1991. The patient's experience of the analyst's subjectivity. In Stephen Mitchell and Lewis Aron, eds., *Relational perspectives: The emergence of a tradition*, 243–268. Hillsdale, NJ: Analytic Press.

Aron, Lewis, and Adrienne Harris, eds. 1993. *The legacy of Sándor Ferenczi*. Hillsdale, NJ: Analytic Press.

Austin, John L. 1962. *How to do things with words*. Oxford: Clarendon.

Bacaltchuck, Josué, and Phillipa Hay. 2003. Antidepressants versus placebo for people with bulimia nervosa. *Cochrane Database of Systematic Reviews* 4, art. no. CD003391.

Bachrach, Leona. 1976. *Deinstitutionalization: An analytical review and sociological perspective*. Washington, DC: Superintendent of Documents, US Government Printing Office.

Baldessarini, Ross. 2001. Drugs and the treatment of psychiatric disorders: Depression and anxiety disorders. In Joel Hardman and Lee Limbird, eds., *Goodman and Gilman's "The pharmacological basis of therapeutics,"* 10th ed., 447–483. New York: McGraw-Hill.

Balint, Michael. 1988. Draft introduction. In *The clinical diary of Sándor Ferenczi*, translated by Michael Balint and Nicola Zarday Jackson, 219–220. Cambridge, MA: Harvard University Press.

Balsam, Rosemary. 2007. Toward less fixed internal transformations of gender commentary on: Melancholy femininity and obsessive-compulsive masculinity: Sex differences in melancholy gender by Meg Jay. *Studies in Gender and Sexuality* 8.2: 137–147.

Barad, Karen. 2007. *Meeting the universe halfway: Quantum physics and the entanglement of matter and meaning*. Durham, NC: Duke University Press.

Barham Carter, A. 1953. The placebo: Its use and abuse. *Lancet* 262.6790: 823.

Baumann, Pierre. 1996. Pharmacokinetic-pharmacodynamic relationship of the selective serotonin reuptake inhibitors. *Clinical Pharmacokinetics* 31.6: 444–460.

Beck, Aaron. 1967. *Depression: Causes and treatments*. Philadelphia: University of Pennsylvania Press.

Beecher, Henry. 1955. The powerful placebo. *Journal of the American Medical Association* 159.17: 1602–1606.

Begley, David. 2003. Understanding and circumventing the blood-brain barrier. *Acta Paediatrica Supplement* 92 .s443: 83–91.

Bell, Gail. 2005. The worried well: The depression epidemic and the medicalisation of our sorrows. *Quarterly Essay* 18. Sydney: Black.

Berger, Douglas, and Isao Fukunishi. 1996. Psychiatric drug development in Japan. *Science*, 273.5273: 318–319.

Berlant, Lauren. 2011. *Cruel optimism*. Durham, NC: Duke University Press.

Berlant, Lauren, and Lee Edelman. 2013. *Sex, or the unbearable*. Durham, NC: Duke University Press.

Bersani, Leo. 1987. Is the rectum a grave? *October* 43: 197–222.

Bersani, Leo. 1990. *The culture of redemption*. Cambridge, MA: Harvard University Press.

Bion, Wilfred. 1959. Attacks on linking. *International Journal of Psychoanalysis* 40: 308–315.

Birke, Lynda. 2000. *Feminism and the biological body*. New Brunswick, NJ: Rutgers University Press.

Birmes, Philippe, Dominique Coppin, Laurent Schmitt, and Dominique Lauque. 2003. Serotonin syndrome: A brief review. *Canadian Medical Association Journal* 168.11: 1439–1442.

Bluhm, Robyn, Anne Jacobson, and Heidi Maibom. 2012. *Neurofeminism: Issues at the intersection of feminist theory and cognitive science*. Houndmills, England: Palgrave Macmillan.

Bonomi, Carlo. 1998. Jones's allegations of Ferenczi's mental deterioration: A reassessment. *International Forum of Psychoanalysis* 7.4: 201–206.

Bordo, Susan. 1993. *Unbearable weight: Feminism, Western culture, and the body*. Berkeley: University of California Press.

Boyer, Edward, and Michael Shannon. 2005. The serotonin syndrome. *New England Journal of Medicine* 352.11: 1112–1120.

Bravo, Javier, Paul Forsythe, Marianne Chew, Emily Escaravage, Hélène Savignac, Timothy Dinan, John Bienenstock, et al. 2011. Ingestion of *Lactobacillus* strain regulates emotional behavior and central GABA receptor expression in a mouse via the vagus nerve. *Proceedings of the National Academy of Sciences* 108.38: 16050–16055.

Breggin, Peter, and Ginger Ross Breggin. 1994. *Talking back to Prozac: What doctors aren't telling you about today's most controversial drug*. New York: St. Martin's.

Brent, David, Graham Emslie, Greg Clarke, Joan Rosenbaum Asarnow, Anthony Spirito, Louise Ritz, Benedetto Vitiello, et al. 2009. Predictors of spontaneous and systematically assessed suicidal adverse events in the treatment of SSRI-resistant depression in adolescents (TORDIA) study. *American Journal of Psychiatry* 166.4: 418–426.

Bridge, Jeffrey, Boris Birmaher, Satish Iyengar, Rémy Barbe, and David Brent. 2009. Placebo response in randomized controlled trials of antidepressants for pediatric major depressive disorder. *American Journal of Psychiatry* 166.1: 42–49.

Bridge, Jeffrey, Satish Iyengar, Cheryl Salary, Rémy Barbe, Boris Birmaher, Harold Alan Pincus, Lulu Ren, et al. 2007. Clinical response and risk for reported suicidal ideation and suicide attempts in pediatric antidepressant treatment: A meta-analysis of randomized controlled trials. *Journal of the American Medical Association* 297.15: 1683–1696.

Brierley, Marjorie. 1942. "Internal objects" and theory. *International Journal of Psycho-analysis* 23: 107–112.

British Medical Journal. 1952. Bottle of medicine. *British Medical Journal* 1.4750: 149–150.

Brockbank, Edward M. 1907. Merycism or rumination in man. *British Medical Journal* 1.2408: 421–427.

Brody, Arthur, Sanjaya Saxena, Paula Stoessel, Laurie Gillies, Lynn Fairbanks, Shervin Alborzian, Michael Phelps, et al. 2001. Regional brain metabolic changes in patients with major depression treated with either paroxetine or interpersonal therapy: Preliminary findings. *Archives of General Psychiatry* 58.7: 631–640.

Brøsen, Kim, and Birgitte Buur Rasmussen. 1996. Selective serotonin re-uptake inhibitors: Pharmacokinetics and drug interactions. In John Feigher and W. F. Boyer, eds., *Selective serotonin re-uptake inhibitors: Advances in basic and clinical practice*, 2nd ed., 87–108. Chichester: John Wiley.

Burton, Robert. 1621/1989. *The anatomy of melancholy*. New York: G. Bell and Sons.

Butler, Judith. 1990. *Gender trouble: Feminism and the subversion of identity*. New York: Routledge.

Butler, Judith. 1994. Sexual traffic. Interview with Judith Butler. *differences: A Journal of Feminist Cultural Studies* 6.2–3: 62–99.

Butler, Judith. 1997. *The psychic life of power: Theories in subjection*. Stanford, CA: Stanford University Press.

Butler, Judith. 1998. Moral sadism and doubting one's own love: Kleinian reflections on melancholia. In Lyndsey Stonebridge and John Phillips, eds., *Reading Melanie Klein*, 179–189. London: Routledge.

Butler, Judith. 2006. Transgender and the spirit of revolt. In Frank Wagner, Kasper König, and Julia Friedrich, eds., *Das achte Feld: Geschlechter, Leben und Begehren in der Kunst seit 1960 / The eighth square: Gender, life, and desire in the visual arts since 1960*, 64–81. Ostfildern, Germany: Hatje Cantz.

Cameron, H. C. 1925. Lumleian Lectures: Some forms of vomiting in infancy. *British Medical Journal* 1.3358: 872–876.

Capasso, Anna, Claudio Petrella, and Walter Milano. 2009. Pharmacological profile of SSRIs and SNRIs in the treatment of eating disorders. *Current Clinical Pharmacology* 4.1: 78–83.

Charney, Dennis, Charles Nemeroff, Lydia Lewis, Sally Laden, Jack Gorman, Eugene Laska, Michael Borenstein, et al. 2002. National Depressive and Manic-Depressive Association consensus statement on the use of placebo in clinical trials of mood disorders. *Archives of General Psychiatry* 59.3: 262–270.

Chen, Mel. 2012. *Animacies: Biopolitics, racial mattering, and queer affect.* Durham, NC: Duke University Press.

Chesler, Phyllis. 1972. *Women and madness: A history of women and the psychiatric profession.* New York: Doubleday.

Chow, Shein-Chung, and Jen-Pei Liu. 2008. *Design and analysis of clinical trials: Concepts and methodologies,* 2nd ed. Hoboken: John Wiley and Sons.

Cixous, Hélène, and Catherine Clément. 1985. The untenable. In Charles Bernheimer and Claire Kahane, eds., *In Dora's Case: Freud—hysteria—feminism,* 276–293. New York: Columbia University Press.

Clouse, R., J. Richter, R. Heading, J. Janssens, and J. Wilson. 1999. Functional esophageal disorders. *Gut* 45.Suppl. 2: 1131–1136.

Committee on Safety of Medicines. 2004. *Report of the CSM expert working group on the safety of selective serotonin reuptake inhibitor antidepressants.* http://www.mhra.gov.uk/home/groups/pl-p/documents/drugsafetymessage/con019472.pdf.

Coole, Diana, and Samantha Frost, eds. 2010. *New materialisms: Ontology, agency, and politics.* Durham, NC: Duke University Press.

Cooper, Melinda. 2008. *Life as surplus: Biotechnology and capitalism in the neoliberal era.* Seattle: University of Washington Press.

Corbett, Ken. 2009a. Boyhood femininity, gender identity disorder, masculine presuppositions, and the anxiety of regulation. *Psychoanalytic Dialogues* 19.4: 353–370.

Corbett, Ken. 2009b. Melancholia and the violent regulation of gender variance: Reply to commentaries. *Psychoanalytic Dialogues* 19.4: 385–392.

Crimp, Douglas. 2002. *Melancholia and moralism: Essays on AIDS and queer politics.* Cambridge, MA: MIT Press.

Cryan, John, and Timothy Dinan. 2014. A light on psychobiotics. *New Scientist* 221.2953: 28–29.

Cvetkovich, Ann. 2007. Public feelings. *South Atlantic Quarterly* 106.3: 459–468.

Cvetkovich, Ann. 2012. *Depression: A public feeling.* Durham, NC: Duke University Press.

Davis, Lennard. 2013. *The end of normal: Identity in a biocultural era.* Ann Arbor: University of Michigan Press.

DeBattista, Charles. 2012. Antidepressant agents. In Bertram Katzung, Susan Masters, and Anthony Trevor, eds., *Basic and clinical pharmacology,* 12th ed. New York: McGraw-Hill.

de Jonghe, Frans, Mariëlle Hendriksen, Gerda van Aalst, Simone Kool, Vjaap Peen, Rien Van, Ellen van den Eijnden, et al. 2004. Psychotherapy alone and combined with pharmacotherapy in the treatment of depression. *British Journal of Psychiatry* 185.1: 37–45.

Delate, Thomas, Alan Gelenberg, Valarie Simmons, and Brenda Motheral. 2004. Trends in the use of antidepressants in a national sample of commercially insured pediatric patients, 1998 to 2002. *Psychiatric Services* 55.4: 387–391.

Derrida, Jacques. 1981. *Dissemination*. Translated by Barbara Johnson. Chicago: University of Chicago Press.

Deutsch, Felix. 1927/1964. Psychoanalysis and internal medicine. In Ralph Kaufman and Marcel Heiman, eds., *Evolution of psychosomatic concepts*, 47–55. New York: International Universities Press.

DeVane, Lindsay. 2009. Principles of pharmacokinetics and pharmacodynamics. In Alan Schatzberg and Charles Nemeroff, eds., *The American psychiatric publishing textbook of psychopharmacology*, 4th ed., 181–199. Arlington, VA: American Psychiatric Publishing.

Dignan, Fiona, Ishaq Abu-Arafeh, and George Russell. 2001. The prognosis of childhood abdominal migraine. *Archive of Diseases in Childhood* 84.5: 415–418.

Dimen, Muriel, and Virginia Goldner. 2002. *Gender and psychoanalytic space: Between clinic and culture*. New York: Other Press.

Dorlan, Newman. 1951. *The American illustrated medical dictionary*, 22nd ed. Philadelphia: W. B. Saunders.

Duman, Ronald. 2002. Pathophysiology of depression: The concept of synaptic plasticity. *European Psychiatry* 2002.17 Suppl. 3: 306–310.

Dunkley, E., Geoffrey Isbister, David Sibbritt, Andrew Dawson, and Ian Whyte. 2003. The Hunter serotonin toxicity criteria: Simple and accurate diagnostic decision rules for serotonin toxicity. *QJM* 96.9: 635–642.

Dupont, Judith. 1988. Introduction. In *The clinical diary of Sándor Ferenczi*, translated by Michael Balint and Nicola Zarday Jackson, xi–xxvii. Cambridge, MA: Harvard University Press.

Dutton, Yulia Chentsova. 2009. Culture and depression. In Rick E. Ingram, ed., *The international encyclopedia of depression*, 194–199. New York: Spring.

Edelman, Lee. 2004. *No future: Queer theory and the death drive*. Durham, NC: Duke University Press.

Elliott, Carl. 2003. *Better than well: American medicine meets the American dream*. New York: Norton.

Elliott, Carl. 2004. Introduction. In Carl Elliott and Tod Chambers, eds., *Prozac as a way of life*. Chapel Hill: University of North Carolina Press.

Emmons, Kimberly. 2010. *Black dogs and blue words: Depression and gender in the age of self-care*. New Brunswick, NJ: Rutgers University Press.

Enck, Paul, Fabrizio Benedetti, and Manfred Schedlowski. 2008. New insights into the placebo and nocebo responses. *Neuron* 59.2: 195–206.

Eng, David. 2000. Melancholia in the late twentieth century. *Signs* 25.4: 1275–1281.

Eng, David, and Shinhee Han. 2006. Desegregating love: Transnational adoption, racial reparation, and racial transitional objects. *Studies in Gender and Sexuality* 7.2: 141–172.

Eng, David, and David Kazanjian. 2003. *Loss: The politics of mourning*. Berkeley: University of California Press.

Epstein, Steven. 2007. *Inclusion: The politics of difference in medical research*. Chicago: University of Chicago Press.

European Medicines Agency. 2005. *European Medicines Agency finalises review of antidepressants in children/adolescents*. http://www.ema.europa.eu/docs/en_GB /document_library/Referrals_document/SSRI_31/WC500013082.pdf. Accessed August 2, 2013.

Fakhoury, Walid, and Stefan Priebe. 2002. The process of deinstitutionalization: An international overview. *Current Opinion in Psychiatry* 15.2: 187–192.

Falzeder, E. 2002. *The complete correspondence of Sigmund Freud and Karl Abraham, 1907–1925*. New York: Karnac.

Farquhar, H. G. 1956. Abdominal migraine in children. *British Medical Journal* 1.4975: 1082–1085.

Fausto-Sterling, Anne. 2000. *Sexing the body: Gender politics and the construction of sexuality*. New York: Basic Books.

Fausto-Sterling, Anne. 2012. *Sex/gender: Biology in a social world*. New York: Routledge.

Fausto-Sterling, Anne, Cynthia Garcia Coll, and Megan Lamarre. 2012a. Sexing the baby: Part 1—What do we really know about sex differentiation in the first three years of life? *Social Science and Medicine* 74.11: 1684–1692.

Fausto-Sterling, Anne, Cynthia Garcia Coll, and Megan Lamarre. 2012b. Sexing the baby: Part 2—Applying dynamic systems theory to the emergences of sex-related differences in infants and toddlers. *Social Science and Medicine* 74.11: 1693–1702.

Ferenczi, Sándor. 1909. Introjection and transference. In *First contributions to psycho-analysis*, translated by Ernest Jones, 35–93. New York: Brunner/Mazel.

Ferenczi, Sándor. 1919. The phenomena of hysterical materialization. In *Further contributions to the theory and techniques of psychoanalysis*, translated by Jane Isabel Suttie, 89–104. New York: Brunner/Mazel.

Ferenczi, Sándor. 1923. "Materialization" in globus hystericus. In *Further contributions to the theory and techniques of psychoanalysis*, translated by Jane Isabel Suttie, 104–105. New York: Brunner/Mazel.

Ferenczi, Sándor. 1924. *Thalassa: A theory of genitality*. Translated by Henry Alden Bunker. New York: Norton.

Ferenczi, Sándor. 1988. *The clinical diary of Sándor Ferenczi*. Edited by Judith Dupont; translated by Michael Balint and Nicola Zarday Jackson. Cambridge, MA: Harvard University Press.

Ferenczi, Sándor, and Sigmund Freud. 1996. *The Correspondence of Sigmund Freud and Sándor Ferenczi*. Vol. 2, 1914–1919. Edited by Ernst Falzeder and Eva Brandt;

translated by Peter T. Hoffer. Cambridge, MA: Belknap Press of Harvard University Press.

Fichter, Manfred, and Karl Pirke. 1990. Endocrine dysfunctions in bulimia (nervosa). In Manfred Fichter, ed., *Bulimia nervosa: Basic research, diagnosis and therapy*, 235–257. Chichester, England: John Wiley.

First, Michael, Allen Frances, and Harold Pincus. 2004. DSM-IV-TR *guidebook*. Washington, DC: American Psychiatric Association.

Fitzgerald, Des, and Felicity Callard. 2015. Social science and neuroscience beyond interdisciplinarity: Experimental entanglements. *Theory, Culture and Society* 32.1: 3–32.

Flückiger, Christoph, A. C. Del Re, Bruce Wampold, Dianne Symonds, and Adam Horvath. 2012. How central is the alliance in psychotherapy? A multilevel longitudinal meta-analysis. *Journal of Counseling Psychology* 59.1: 10–17.

Fluoxetine Bulimia Nervosa Research Group. 1992. Fluoxetine on the treatment of bulimia nervosa. *Archives of General Psychiatry* 49: 139–147.

Fonagy, Peter. 2010. The changing shape of clinical practice: Driven by science or by pragmatics? *Psychoanalytic Psychotherapy* 24.1: 22–43.

Fonagy, Peter, Mary Target, George Gergely, Jon Allen, and Anthony Bateman. 2003. The developmental roots of borderline personality disorder in early attachment relationships: A theory and some evidence. *Psychoanalytic Inquiry* 23.3: 412–459.

Foucault, Michel. 1978. *The history of sexuality: Volume 1*. Harmondsworth, England: Penguin.

Franco, Kathleen, Nancy Campbell, Marijo Tamburrino, and Cynthia Evans. 1993. Rumination: The eating disorder of infancy. *Child Psychiatry and Human Development* 24.2: 91–97.

Frankel, Lois. 1991. *Women, anger and depression: Strategies for self-empowerment*. Deerfield Beach, FL: Health Communications.

Franklin, Sarah. 2007. *Dolly mixtures: The remaking of genealogy*. Durham, NC: Duke University Press.

Franklin, Sarah. 2013. *Biological relatives: IVF, stem cells, and the future of kinship*. Durham, NC: Duke University Press.

Fraser, Mariam. 2001. The nature of Prozac. *History of the Human Sciences* 14.3: 56–84.

Fraser, Mariam. 2003. Material theory: Duration and the serotonin hypothesis of depression. *Theory, Culture and Society* 20.5: 1–26.

Fraser, Mariam. 2009. Standards, populations, and difference. *Cultural Critique* 71: 47–80.

Freud, Sigmund. 1891. *On aphasia: A critical study*. Translated by E. Stengel. New York: International Universities Press.

Freud, Sigmund. 1892. Draft G: Melancholia. In James Strachey, ed. and trans., *The standard edition of the complete psychological works of Sigmund Freud*, Vol. 1: Pre-psychoanalytic publications and unpublished drafts, 200–206. London: The Hogarth Press and the Institute of Psychoanalysis.

Freud, Sigmund. 1893a. Some points for a comparative study of organic and hysterical motor paralyses. In James Strachey, ed. and trans., *The standard edition of the complete psychological works of Sigmund Freud*, Vol. 1: Pre-psychoanalytic publications and unpublished drafts, 160–172. London: The Hogarth Press and the Institute of Psychoanalysis.

Freud, Sigmund. 1893b. The psychotherapy of hysteria. In James Strachey, ed. and trans., *The standard edition of the complete psychological works of Sigmund Freud*, Vol. 2: 1893–1895: Studies on hysteria, 253–305. London: The Hogarth Press and the Institute of Psychoanalysis.

Freud, Sigmund. 1900. The interpretation of dreams [first part 1900; second part 1900–1901]. In James Strachey, ed. and trans., *The standard edition of the complete psychological works of Sigmund Freud*, Vols. 4 and 5: 1900–1901. London: The Hogarth Press and the Institute of Psychoanalysis.

Freud, Sigmund. 1905. Fragment of an analysis of a case of hysteria. In James Strachey, ed. and trans., *Standard edition of the complete psychological works of Sigmund Freud*, Vol. 7: 1901–1905: A Case of Hysteria, Three Essays on Sexuality and Other Works, 3–122. London: The Hogarth Press and the Institute of Psychoanalysis.

Freud, Sigmund. 1909a. Analysis of a phobia in a five-year-old boy [Little Hans]. In James Strachey, ed. and trans., *The standard edition of the complete psychological works of Sigmund Freud*, Vol. 10: 1909: Two case histories ("Little Hans" and the "Rat Man"), 3–149. London: The Hogarth Press and the Institute of Psychoanalysis.

Freud, Sigmund. 1909b. Notes upon a case of obsessional neurosis [Rat Man]. In James Strachey, ed. and trans., *The standard edition of the complete psychological works of Sigmund Freud*, Vol. 10: 1909: Two Case Histories ("Little Hans" and the "Rat Man"), 153–318. London: The Hogarth Press and the Institute of Psychoanalysis.

Freud, Sigmund. 1915. The unconscious. In James Strachey, ed. and trans., *The standard edition of the complete psychological works of Sigmund Freud*, Vol. 14: 1914–1916: On the history of the psycho-analytic movement, Papers on metapsychology, and other works, 159–215. London: The Hogarth Press and the Institute of Psychoanalysis.

Freud, Sigmund. 1917a. Mourning and melancholia. In James Strachey, ed. and trans., *The standard edition of the complete psychological works of Sigmund Freud*, Vol. 14: 1914–1916: On the history of the psycho-analytic movement, Papers on metapsychology, and other works, 237–258. London: The Hogarth Press and the Institute of Psychoanalysis.

Freud, Sigmund. 1917b. The paths to the formation of symptoms [Introductory Lecture 23]. In James Strachey, ed. and trans., *The standard edition of the complete psychological works of Sigmund Freud*, Vol. 16: 1916–1917: Introductory lectures on *psycho-analysis (part 3)*, 358–377. London: The Hogarth Press and the Institute of Psychoanalysis.

Freud, Sigmund. 1920. Beyond the pleasure principle. In James Strachey, ed. and trans., *The standard edition of the complete psychological works of Sigmund Freud*, Vol. 18: 1920–1922: *Beyond the pleasure principle, Group psychology, and other works*, 1–64. London: The Hogarth Press and the Institute of Psychoanalysis.

Freud, Sigmund. 1931. Female sexuality. In James Strachey, ed. and trans., *The standard edition of the complete psychological works of Sigmund Freud*, Vol. 21: 1927–1931: *The Future of an illusion, Civilization and its discontents, and other works*, 221–244. London: The Hogarth Press and the Institute of Psychoanalysis.

Freud, Sigmund. 1939. Moses and monotheism: Three essays. In James Strachey, ed. and trans., *The standard edition of the complete psychological works of Sigmund Freud*, Vol. 23: 1937–1939: *Moses and monotheism, An outline of psycho-analysis, and other works*, 3–137. London: The Hogarth Press and the Institute of Psychoanalysis.

Gallop, Jane. 1988. *Thinking through the body*. New York: Columbia University Press.

Gardiner, Judith Kegan. 1995. Review: Can Ms. Prozac talk back? Feminism, drugs, and social constructionism. *Feminist Studies* 21.3: 501–517.

Gardner, Paula. 2003. Distorted packaging: Marketing depression as illness, drugs as cure. *Journal of Medical Humanities* 24.1–2: 105–130.

Garland, E. Jane, Stan Kutcher, and Adil Virani. 2009. 2008 position paper on using SSRIs in children and adolescents. *Journal of the Canadian Academy of Child and Adolescent Psychiatry* 18.2: 160–165.

Geffen, Nathan. 1966. Rumination in man: Report of a case. *American Journal of Digestive Diseases* 11.12: 963–972.

Gibbons, Robert, Hendricks Brown, Kwan Hur, Sue Marcus, Dulal Bhaumik, Joëlle Erkens, Ron Herings, et al. 2007. Early evidence on the effects of regulators' suicidality warnings on SSRI prescriptions and suicide in children and adolescents. *American Journal of Psychiatry* 164.9: 1356–1363.

Giffney, Noreen. 2008. Queer apocal(o)ptic/ism: The death drive and the human. In Noreen Giffney and Myra Hird, eds., *Queering the non/human*, 55–78. Farnham, Surrey, England: Ashgate.

Giffney, Noreen, and Myra Hird. 2008. *Queering the non/human*. Farnham, Surrey, England: Ashgate.

Glannon, Walter. 2002. The psychology and physiology of depression. *Philosophy, Psychiatry, and Psychology* 9.3: 265–269.

Goldstein, David, Michael Wilson, Richard Ascroft, and Mahir Al-Banna. 1999. Effectiveness of fluoxetine therapy in bulimia nervosa regardless of comorbid depression. *International Journal of Eating Disorders* 25.1: 19–27.

Goldstein, David, Michael Wilson, Vicki Thompson, Janet Potvin, Alvin Rampey, and the Fluoxetine Bulimia Nervosa Research Group. 1995. Long-term fluoxetine treatment of bulimia nervosa. *British Journal of Psychiatry* 166.5: 660–666.

Gray, Henry. 1918. *Anatomy of the human body*, 20th ed. Philadelphia: Lea and Febiger.

Griggers, Camilla. 1997. *Becoming woman*. Minneapolis: University of Minnesota Press.

Griggers, Camilla. 1998. The micropolitics of biopsychiatry. In Margrit Shildrik and Janet Price, eds., *Vital signs: Feminist reconfigurations of the bio/logical body*, 132–144. Edinburgh: Edinburgh University Press.

Grosz, Elizabeth. 2004. *The nick of time: Politics, evolution, and the untimely*. Durham, NC: Duke University Press.

Grosz, Elizabeth. 2005. *Time travels: Feminism, nature, power*. Durham, NC: Duke University Press.

Halberstam, Judith. 2006. The politics of negativity in recent queer theory. PMLA 121.3: 823–825.

Halley, Ian. 2004. Queer theory by men. *Duke Journal of Gender, Law, and Policy* 11: 7–53.

Halley, Janet. 2006. *Split decisions: How and why to take a break from feminism*. Princeton, NJ: Princeton University Press.

Hamilton, Max. 1960. A rating scale for depression. *Journal of Neurology, Neurosurgery, and Psychiatry* 23.1: 56–62.

Hammad, Tarek, Thomas Laughren, and Judith Racoosin. 2006. Suicidality in pediatric patients treated with antidepressant drugs. *Archives of General Psychiatry* 63.3: 332–339.

Handfield-Jones, R. P. C. 1953. A bottle of medicine from the doctor. *Lancet* 262.6790: 823–825.

Hanson, Ellis. 2011. The future's Eve: Reparative reading after Sedgwick. *South Atlantic Quarterly* 110.1: 101–119.

Haraway, Donna J. 2003. *The companion species manifesto: Dogs, people, and significant otherness*. Chicago: Prickly Paradigm.

Haraway, Donna J. 2007. *When species meet*. Minneapolis: University of Minnesota Press.

Harrington, Anne, ed. 1997. *The placebo effect: An interdisciplinary exploration*. Cambridge, MA: Harvard University Press.

Harrington, Anne. 2006. The many meanings of the placebo effect: Where they came from, why they matter. *BioSocieties* 1.2: 181–193.

Harrington, R., Michael Rutter, and Eric Fombonne. 1996. Developmental pathways in depression: Multiple meanings, antecedents, and endpoints. *Development and Psychopathology* 8: 601–616.

Haynal, André. 2002. *Disappearing and reviving: Sándor Ferenczi in the history of psychoanalysis*. London: Karnac.

Healy, David. 1997. *The antidepressant era*. Cambridge, MA: Harvard University Press.

Healy, David. 2004. *Let them eat Prozac: The unhealthy relationship between the pharmaceutical industry and depression*. New York: New York University Press.

Healy, David, and Chris Whitaker. 2003. Antidepressants and suicide: Risk-benefit conundrums. *Journal of Psychiatry and Neuroscience* 28.5: 331–337.

Hekman, Susan. 2010. *The material of knowledge: Feminist disclosures.* Bloomington: Indiana University Press.

Hernández, María, and Appu Rathinavelu. 2006. *Basic pharmacology: Understanding drug actions and reactions.* Boca Raton, FL: CRC Press / Taylor and Francis.

Hiemke, Christoph, and Sebastian Härtter. 2000. Pharmacokinetics of selective serotonin reuptake inhibitors. *Pharmacology and Therapeutics* 85.1: 11–28.

Hinshelwood, Robert. 1991. *A dictionary of Kleinian thought.* Northvale, NJ: Jason Aronson.

Hinshelwood, Robert, Susan Robinson, and Oscar Zarate. 1997. *Introducing Melanie Klein: A graphic guide.* London: Icon.

Hippocrates. 1978. *Hippocratic writings.* Edited by G. Lloyd; translated by John Chadwick, William Mann, I. Lonie, and E. Withington. Harmondsworth, England: Penguin.

Hird, Myra. 2009. *The origins of sociable life: Evolution after science studies.* Houndsmills, England: Palgrave Macmillan.

Hirshbein, Laura. 2009. *American melancholy: Constructions of depression in the twentieth century.* New Brunswick, NJ: Rutgers University Press.

Holsboer, Florian. 2008. How can we realize the promise of personalized antidepressant medicines? *Nature Reviews Neuroscience* 9.8: 638–646.

Horder, Jamie, Paul Matthews, and Robert Waldmann. 2011. Placebo, Prozac and PLoS: Significant lessons for psychopharmacology. *Journal of Psychopharmacology* 25.10: 1277–1288.

Hornbacher, Marya. 1998. *Wasted: A memoir of anorexia and bulimia.* New York: Harper Flamingo.

Horvath, Adam, A. C. Del Re, Christoph Flückiger, and Dianne Symonds. 2011. Alliance in individual psychotherapy. *Psychotherapy* 48.1: 9–16.

Horwitz, Allan, and Jerome Wakefield. 2007. *The loss of sadness: How psychiatry transformed normal sorrow into depressive disorder.* Oxford: Oxford University Press.

Hsu, L. K. G., Ross S. Kalucy, Arthur H. Crisp, J. Koval, C. N. Chen, M. E. Carruthers, and K. J. Zilkha. 1977. Early morning migraine: Nocturnal plasma levels of catecholamines, tryptophan, glucose, and free fatty acids and sleep encephalographs. *Lancet* 309.8009: 447–451.

Hughes, Patrick, Lloyd Wells, Carol Cunningham, and Duane Ilstrup. 1986. Treating bulimia with desipramine: A double-blind, placebo-controlled study. *Archives of General Psychiatry* 43.2: 182–186.

Hunter, Aimee, Andrew Leuchter, Melinda Morgan, and Ian Cook. 2006. Changes in brain function (quantitative EEG cordance) during placebo lead-in and treatment outcomes in clinical trials for major depression. *American Journal of Psychiatry* 163.8: 1426–1432.

Hyman, Paul, Peter Milla, Marc Benninga, Geoff Davidson, David Fleisher, and Jan Taminiau. 2006. Childhood functional gastrointestinal disorders: Neonate/toddler. *Gastroenterology* 130.5: 1519–1526.

Ingram, Rick. 2009. *The international encyclopedia of depression.* New York: Springer.

Isaacs, Susan. 1948. The nature and function of phantasy. *International Journal of Psychoanalysis* 29: 73–97.

Jackson, Stanley. 1986. *Melancholia and depression: From Hippocratic times to modern times.* New Haven, CT: Yale University Press.

Jantzen, Gwen, and Joseph Robinson. 2002. Sustained- and controlled-release drug-delivery systems. In Gilbert Banker and Christopher Rhodes, eds., *Modern pharmaceutics*, 4th ed., revised and expanded, 501–528. New York: Marcel Dekker.

Johnson, Barbara. 1981. Translator's introduction. In Jacques Derrida, *Dissemination*, vii–xxxiii. Chicago: University of Chicago Press.

Jones, Ernest. 1955. *The life and work of Sigmund Freud. Vol. 2: Years of maturity, 1901–1919.* New York: Basic Books.

Jones, Ernest. 1957. *The life and work of Sigmund Freud. Vol. 3: The last phase, 1919–1939.* New York: Basic Books.

Jones, Ernest, and Sigmund Freud. 1993. *The complete correspondence of Sigmund Freud and Ernest Jones, 1908–1939.* Edited by R. Andrew Paskauskas. London: Karnac.

Jordan-Young, Rebecca. 2010. *Brain storm: The flaws in the science of sex differences.* Cambridge, MA: Harvard University Press.

Jurist, Elliot. 2010. Elliot Jurist interviews Peter Fonagy. *Psychoanalytic Psychology* 27.1: 2–7.

Kaptchuk, Ted. 1998. Powerful placebo: The dark side of the randomised controlled trial. *Lancet* 351.9117: 1722–1725.

Kaptchuk, Ted, John Kelley, Aaron Deykin, Peter Wayne, Louis Lasagna, Ingrid Epstein, Irving Kirsch, et al. 2008. Do "placebo responders" exist? *Contemporary Clinical Trials* 29.4: 587–595.

Katzung, Bertram. 2012. *Basic and clinical pharmacology.* New York: McGraw-Hill.

Keller, Evelyn Fox. 2000. *Century of the gene.* Cambridge, MA: Harvard University Press.

Keller, Evelyn Fox. 2002. *Making sense of life: Explaining biological development with models, metaphors, and machines.* Cambridge, MA: Harvard University Press.

Keller, Evelyn Fox. 2010. *The mirage of a space between nature and nurture.* Durham, NC: Duke University Press.

Keller, Martin, James McCullough, Daniel Klein, Bruce Arnow, David L. Dunner, Alan J. Gelenberg, John C. Markowitz, et al. 2000. A comparison of nefazodone, the cognitive behavioral-analysis system of psychotherapy, and their combination for the treatment of chronic depression. *New England Journal of Medicine* 342.20: 1462–1470.

Khan, Arif, Amritha Bhat, Russell Kolts, Michael Thase, and Walter Brown. 2010. Why has the antidepressant–placebo difference in antidepressant clinical trials diminished over the past three decades? CNS *Neuroscience and Therapeutics* 16.4: 217–226.

Khan, Seema, Paul Hyman, Jose Cocjin, and Carlo di Lorenzo. 2000. Rumination syndrome in adolescents. *Journal of Pediatrics* 136.4: 528–531.

King, Pearl, and Riccardo Steiner, eds. 1991. *The Freud-Klein controversies, 1941–45.* London: Routledge.

King v. McInerney. 2009. Complaint for wrongful death. Case no. 56-2009-00337175-CU-PP-VTA. Superior Court of the State of California.

Kipnis, Laura. 2006. Response to "The traffic in women." *Women's Studies Quarterly* 34.1–2: 434–437.

Kirby, Vicki. 1997. *Telling flesh: The substance of the corporeal.* New York: Routledge.

Kirby, Vicki. 2011. *Quantum anthropologies: Life at large.* Durham, NC: Duke University Press.

Kirmayer, Laurence. 2002. Psychopharmacology in a globalizing world: The use of antidepressants in Japan. *Transcultural Psychiatry* 39.3: 295–322.

Kirsch, Irving. 2010. *The emperor's new drugs: Exploding the antidepressant myth.* New York: Basic Books.

Kirsch, Irving, Brett Deacon, Tania Huedo-Medina, Alan Scoboria, Thomas Moore, and Blair Johnson. 2008. Initial severity and antidepressant benefits: A meta-analysis of data submitted to the Food and Drug Administration. *PLoS Medicine* 5.2: e45.

Kirsch, Irving, Thomas Moore, Alan Scoboria, and Sarah Nicholls. 2002. The emperor's new drugs: An analysis of antidepressant medication data submitted to the US Food and Drug Administration. *Prevention and Treatment* 5.1: 23a.

Kirsch, Irving, and Guy Sapirstein. 1998. Listening to Prozac but hearing placebo: A meta-analysis of antidepressant medication. *Prevention and Treatment* 1.2: n.p. Available at http://psycnet.apa.org/journals/pre/1/2/.

Klein, Daniel, Joseph Schwartz, Neil Santiago, Dina Vivian, Carina Vocisano, Louis Castonguay, Bruce Arnow, et al. 2003. Therapeutic alliance in depression treatment: Controlling for prior change and patient characteristics. *Journal of Consulting and Clinical Psychology* 71.6: 997–1006.

Klein, Donald. 1998. Listening to meta-analysis, but hearing bias. *Prevention and Treatment* 1.2: 6c. Available at http://psycnet.apa.org/journals/pre/1/2/.

Klein, Melanie. 1935/1975. A contribution to the psychogenesis of manic-depressive states. In *Love, guilt, and reparation, and other works, 1921–1945,* 262–289. New York: Free Press.

Klein, Melanie. 1936/1975. Weaning. In *Love, guilt, and reparation, and other works, 1921–1945,* 290–305. New York: Free Press.

Klein, Melanie. 1940/1975. Mourning and its relation to manic-depressive states. In *Love, guilt, and reparation, and other works, 1921–1945*, 344–369. New York: Free Press.

Kleinman, Arthur. 1986. *Social origins of distress and disease: Neurasthenia, depression, and pain in modern China*. New Haven, CT: Yale University Press.

Kramer, Peter. 1993. *Listening to Prozac: A psychiatrist explores antidepressant drugs and the remaking of the self*. New York: Penguin.

Kring, Ann, Sheri Johnson, Gerald C. Davison, and John M. Neale. 2010. *Abnormal psychology*, 11th ed. Hoboken: John Wiley.

Kristeva, Julia. 1989. *Black sun: Depression and melancholia*. Translated by Leon Roudiez. New York: Columbia University Press.

Krupnick, Janice, Stuart Sotsky, Sam Simmens, Janet Moyer, Irene Elkin, John Watkins, and Paul Pilkonis. 1996. The role of the therapeutic alliance in psychotherapy and pharmacotherapy outcome: Findings in the National Institute of Mental Health Treatment of Depression Collaborative Research Program. *Journal of Consulting and Clinical Psychology* 64.3: 532–539.

Kunzel, Regina. 2011. Queer studies in queer times: Conference review of "Rethinking Sex," University of Pennsylvania, March 4–6, 2009. *GLQ* 17.1: 155–165.

Lagassé, Paul, ed. 2000. *The Columbia encyclopedia*, 6th ed. New York: Columbia University Press.

Lamb, Richard. 1998. Deinstitutionalization at the beginning of the new millennium. *Harvard Review of Psychiatry* 61:1–10.

Lancet. 2004. Depressing research (Editorial). *Lancet* 363.9418: 1335.

Landecker, Hannah. 2010. *Culturing life: How cells became technologies*. Cambridge, MA: Harvard University Press.

Landecker, Hannah. 2013. The metabolism of philosophy, in three parts. In Bernhard Malkmus and Ian Cooper, eds., *Dialectic and paradox: Configurations of the third in modernity*, 193–224. Bern: Peter Lang AG.

Laplanche, Jean, and Jean-Bertrand Pontalis. 1988. *The language of psychoanalysis*. London: Karnac.

Lasagna, Louis, Frederick Mosteller, John von Felsinger, and Henry Beecher. 1954. A study of placebo response. *American Journal of Medicine* 16.6: 770–779.

Leader, Darian. 2008. *The new black: Mourning, melancholia and depression*. London: Penguin.

Leibowitz, Sarah. 1990. The role of serotonin in eating disorders. *Drugs Suppl.* 3: 33–48.

Leombruni, Paolo, Federico Amianto, Nadia Delsedime, Carla Gramaglia, Giovanni Abbate-Daga, and Secondo Fassino. 2006. Citalopram versus fluoxetine for the treatment of patients with bulimia nervosa: A single-blind randomized controlled trial. *Advances in Therapy* 23.3: 481–494.

Leonard, Brian. 1996. The comparative pharmacological properties of selective serotonin re-uptake inhibitors in animals. In J. P. Feigher and W. F. Boyer, eds., *Selective serotonin re-uptake inhibitors: Advances in basic and clinical practice*, 2nd ed., 35–62. Chichester: John Wiley and Sons.

Leslie, Laurel, Thomas Newman, Joan Chesney, and James Perrin. 2005. The Food and Drug Administration's deliberations on antidepressant use in pediatric patients. *Pediatrics* 116.1: 195–204.

Leuchter, Andrew, Ian Cook, Elise Witte, Melinda Morgan, and Michelle Abrams. 2002. Changes in brain function of depressed subjects during treatment with placebo. *American Journal of Psychiatry* 159.1: 122–129.

Levine, Jon, Newton Gordon, and Howard Fields. 1978. The mechanism of placebo analgesia. *Lancet* 312.8091: 654–657.

Lewis, Bradley. 2006. *Moving beyond Prozac, DSM, and the new psychiatry: The birth of postpsychiatry.* Ann Arbor: University of Michigan Press.

Liebert, Rachel, and Nicola Gavey. 2009. "There are always two sides to these things": Managing the dilemma of serious adverse effects from SSRIs. *Social Science and Medicine* 68.10: 1882–1891.

Likierman, Meira. 2001. *Melanie Klein: Her work in context.* London: Continuum.

Littlefield, Melissa, and Jenell Johnson, eds. 2012. *The neuroscientific turn: Transdisciplinarity in the age of the brain.* Ann Arbor: University of Michigan Press.

Litvak, Joseph. 1997. *Strange gourmets: Sophistication, theory, and the novel.* Durham, NC: Duke University Press.

Love, Heather. 2010. Truth and consequences: On paranoid reading and reparative reading. *Criticism* 52.2: 235–241.

Malabou, Catherine. 2008. *What should we do with our brain?* New York: Fordham University Press.

Martin, Daniel, John Garske, and Katherine Davis. 2000. Relation of the therapeutic alliance with outcome and other variables: A meta-analytic review. *Journal of Consulting and Clinical Psychology* 68.3: 438–450.

Martin, Emily. 1987. *The woman in the body: A cultural analysis of reproduction.* Boston: Beacon.

McIvor, David. 2012. Bringing ourselves to grief: Judith Butler and the politics of mourning. *Political Theory* 40.4: 409–436.

McManus, Peter, Andrea Mant, Philip Mitchell, Helena Britt, and John Dudley. 2003. Use of antidepressants by general practitioners and psychiatrists in Australia. *Australian and New Zealand Journal of Psychiatry* 37.2: 184–189.

Menking, Manfred, John Wagnitz, Josef Burton, Dean Coddington, and Juan Sotos. 1969. Rumination—a near fatal psychiatric disease of infancy. *New England Journal of Medicine* 280.15: 802–804.

Metzl, Jonathan. 2003. *Prozac on the couch: Prescribing gender in the era of wonder drugs.* Durham, NC: Duke University Press.

Meyers, Todd. 2013. *The clinic and elsewhere: Addiction, adolescents, and the afterlife of therapy*. Seattle: University of Washington Press.

Milla, Peter, Paul Hyman, Marc Benninga, Geoffrey Davidson, David Fleischer, and Jan Taminiau. 2006. Childhood functional gastrointestinal disorders: Neonate/Toddler. In Douglas Drossman, ed., *Rome III: The functional gastrointestinal disorders*, 3rd ed., 687–722. McLean: Degnon Associates.

Miller, Anita. 2007. Social neuroscience of child and adolescent depression. *Brain and Cognition* 65.1: 47–68.

Mitchell, James, Linda Fletcher, Karen Hanson, Melissa Pederson Mussell, Harold Seim, Ross Crosby, and Mahir Al-Banna. 2001. The relative efficacy of fluoxetine and manual-based self-help in the treatment of outpatients with bulimia nervosa. *Journal of Clinical Psychopharmacology* 21.3: 298–304.

Mol, Annemarie. 2002. *The body multiple: Ontology in medical practice*. Durham, NC: Duke University Press.

Montagne, Michael. 1996. The pharmakon phenomenon: Cultural conceptions of drugs and drug use. In Peter Davis, ed., *Contested ground: Public purpose and private interest in the regulation of prescription drugs*, 11–25. New York: Oxford University Press.

Morgen, Sandra. 2002. *Into our own hands: The women's health movement in the United States, 1969–1990*. New Brunswick, NJ: Rutgers University Press.

Mortimer-Sandilands, Catriona, and Bruce Erickson. 2010. *Queer ecologies: Sex, nature, politics, desire*. Bloomington: Indiana University Press.

Murphy, Michelle. 2012. *Seizing the means of reproduction: Entanglements of feminism, health, and technoscience*. Durham, NC: Duke University Press.

Murray, M. L., Corinne de Vries, and I. C. K. Wong. 2004. A drug utilisation study of antidepressants in children and adolescents using the General Practice Research Database. *Archives of Disease in Childhood* 89.12: 1098–1102.

NICE (National Institute for Health and Clinical Excellence). 2005. *Depression in children and young people: Identification and management in primary, community and secondary care*. September. http://www.nice.org.uk/guidance/cg28.

Nolen-Hoeksema, Susan. 1990. *Sex differences in depression*. Stanford, CA: Stanford University Press.

Oates, John, and Albert Sjoerdsma. 1960. Neurologic effects of tryptophan in patients receiving a monoamine oxidase inhibitor. *Neurology* 10.12: 1076–1078.

O'Brien, Michael, Barbara Bruce, and Michael Camelleri. 1995. The rumination syndrome: Clinical features rather than manometric diagnosis. *Gastroenterology* 108.4: 1024–1029.

Ogden, Thomas. 1994. The analytic third: Working with intersubjective clinical facts. *International Journal of Psychoanalysis* 75.1: 3–20.

Ogden, Thomas. 1999. Afterword. In Stephen Mitchell and Lewis Aron, eds., *Relational psychoanalysis: The emergence of a tradition*, 487–492. Hillsdale, NJ: Analytic Press.

Olden, Kevin. 2001. Rumination. *Current Treatment Options in Gastroenterology* 4.4: 351–358.

Olsen, Richard, and Guo-Dong Li. 2012. GABA. In Scott Brady, George Siegel, Wayne Albers, and Donald Price, eds., *Basic neurochemistry principles of molecular, cellular and medical neurobiology*, 8th ed., 367–376. Waltham, MA: Elsevier.

Oyama, Susan. 2000. *The ontogeny of information: Developmental systems and evolution.* Durham, NC: Duke University Press.

Pampallona, Sandro, Paolo Bollini, Giuseppe Tibaldi, Bruce Kupelnick, and Carmine Munizza. 2004. Combined pharmacotherapy and psychological treatment for depression: A systematic review. *Archives of General Psychiatry* 61.7: 714–719.

Parry-Jones, Brenda. 1994. Merycism or rumination disorder. A historical investigation and current assessment. *British Journal of Psychiatry* 165.3: 303–314.

Pellow, Sharon, Philippe Chopin, Sandra E. File, and Mike Briley. 1985. Validation of open: closed arm entries in an elevated plus-maze as a measure of anxiety in the rat. *Journal of Neuroscience Methods* 14.3: 149–167.

Pepper, Oliver. 1945. A note on the placebo. *American Journal of Pharmacy* 117: 409–412.

Persson, Asha. 2004. Incorporating *pharmakon:* HIV, medicine, and body shape change. *Body and Society* 10.4: 45–67.

Petryna, Adriana. 2009. *When experiments travel: Clinical trials and the global search for human subjects.* Princeton, NJ: Princeton University Press.

Petryna, Adriana, Andrew Lakoff, and Arthur Kleinman, eds. 2006. *Global pharmaceuticals: Ethics, markets, practices.* Durham, NC: Duke University Press.

Piaget, Jean. 1929/1997. *The child's conception of the world.* London: Routledge.

Pope, Harrison, and James Hudson. 1986. Antidepressant drug therapy of bulimia: Current status. *Journal of Clinical Psychiatry* 47.7: 339–345.

Pope, Harrison, James Hudson, Jeffrey Jonas, and Deborah Yurgelun-Todd. 1983. Bulimia treated with imipramine: A placebo-controlled, double-blind study. *American Journal of Psychiatry* 140.5: 554–558.

Potter, William, and Leo Hollister. 2001. Antidepressant agents. In Bertram Katzung, ed., *Basic and clinical pharmacology*, 8th ed., 498–511. New York: McGraw-Hill.

Rachman, Arnold. 1997. *Sandor Ferenczi: The psychoanalyst of tenderness and passion.* Lanham, MD: Jason Aronson.

Radden, Jennifer, ed. 2000. *The nature of melancholy: From Aristotle to Kristeva.* Oxford: Oxford University Press.

Radden, Jennifer. 2003. Is this dame melancholy? Equating today's depression and past melancholia. *Philosophy, Psychiatry, and Psychology* 10.1: 37–52.

Rasquin-Weber, Andree, P. E. Hyman, Salvatore Cucchiara, D. R. Fleisher, J. S. Hyams, P. J. Milla, and A. Staiano. 1999. Childhood functional gastrointestinal disorders. *Gut* 45.Suppl. II: 1160–1168.

Rentoul, Robert. 2010. *Ferenczi's language of tenderness: Working with disturbances from the earliest years.* Lanham, MD: Jason Aronson.

Richardson, Sarah. 2013. *Sex itself: The search for male and female in the human genome.* Chicago: University of Chicago Press.

Rief, Winfried, Yvonne Nestoriuc, Sarah Weiss, Eva Welzel, Arthur Barsky, and Stefan Hofmann. 2009. Meta-analysis of the placebo response in antidepressant trials. *Journal of Affective Disorders* 118.1: 1–8.

Ritschel, Wolfgang, and Gregory Kearns. 2004. *Handbook of basic pharmacokinetics,* 6th ed. Washington, DC: American Pharmacists Association.

Roberts, Celia. 2007. *Messengers of sex: Hormones, biomedicine and feminism.* Cambridge: Cambridge University Press.

Robinson, Paul, and Letizia Grossi. 1986. Gag reflex in bulimia nervosa. *Lancet* 328.8500: 221.

Rose, Jacqueline. 1993. *Why war? Psychoanalysis, politics and the return to Melanie Klein.* Oxford: Blackwell.

Rose, Nikolas. 2003. Neurochemical selves. *Society* 41.1: 46–59.

Rose, Nikolas. 2004. Becoming neurochemical selves. In Nico Stehr, ed., *Biotechnology: Between commerce and civil society,* 89–126. New Brunswick, NJ: Transaction.

Rose, Nikolas. 2007. *The politics of life itself: Biomedicine, power, and subjectivity in the twenty-first century.* Princeton, NJ: Princeton University Press.

Rosengarten, Marsha. 2009. *HIV interventions: Biomedicine and the traffic between information and flesh.* Seattle: University of Washington Press.

Rubin, Gayle. 1975. The traffic in women: Notes on the "political economy" of sex. In Rayna Reiter, ed., *Toward an anthropology of women,* 157–210. New York: Monthly Review Press.

Rubin, Gayle. 1984. Thinking sex: Notes for a radical theory of the politics of sexuality. In Carole Vance, ed., *Pleasure and danger: Exploring female sexuality,* 267–319. Boston: Routledge.

Rubin, Gayle. 1994. Sexual traffic. Interview with Judith Butler. *differences: A Journal of Feminist Cultural Studies* 6.2–3: 62–99.

Rudnytsky, Peter, Antal Bókay, and Patrizia Giampieri-Deutsch, eds. 2000. *Ferenczi's turn in psychoanalysis.* New York: New York University Press.

Russell, Gerald. 1979. Bulimia nervosa: An ominous variant of anorexia nervosa. *Psychological Medicine* 9.3: 429–448.

Russell, Gerald. 1990. The diagnostic status and clinical assessment of bulimia nervosa. In Manfred M. Fichter, ed., *Bulimia nervosa: Basic research, diagnosis, and therapy,* 17–36. New York: John Wiley.

Rutherford, Bret, and Steven Roose. 2013. A model of placebo response in antidepressant clinical trials. *American Journal of Psychiatry* 170.7: 723–733.

Salamon, Gayle. 2007. Melancholia, ambivalent presence and the cost of gender. Commentary on paper by Meg Jay. *Studies in Gender and Sexuality* 8.2: 149–164.

Salamon, Gayle. 2009. Humiliation and transgender regulation: Commentary on paper by Ken Corbett. *Psychoanalytic Dialogues* 19.4: 376–384.

Salamone, John. 2002. Antidepressants and placebos: Conceptual problems and research strategies. *Prevention and Treatment* 5.1: n.p. Article 24. Available at http://psycnet.apa.org/journals//pre/5/1/.

Sánchez-Pardo, Esther. 2003. *Cultures of the death drive: Melanie Klein and modernist melancholia.* Durham, NC: Duke University Press.

Sartorius, Norman, Thomas Baghai, David Baldwin, Barbara Barrett, Ursula Brand, Wolfgang Fleischhacker, Guy Goodwin, et al. 2007. Antidepressant medications and other treatments of depressive disorders: A CINP Task Force report based on a review of evidence. *International Journal of Neuropsychopharmacology* 10.1: 1–207.

Schatzberg, Alan, and Charles Nemeroff, eds. 2009. *Essentials of clinical psychopharmacology,* 4th ed. Arlington, VA: American Psychiatric Publishing.

Schiesari, Juliana. 1992. *The gendering of melancholia: Feminism, psychoanalysis, and the symbolics of loss in Renaissance literature.* Ithaca, NY: Cornell University Press.

Schore, Allan. 1994. *Affect regulation and the origin of the self: The neurobiology of emotional development.* Hillsdale, NJ: Lawrence Erlbaum.

Schuske, Kim, Asim Beg, and Erik Jorgensen. 2004. The GABA nervous system in *C. elegans. Trends in Neurosciences* 27.7: 407–414.

Sedgwick, Eve Kosofsky. 1997. Paranoid reading and reparative reading; or, you're so paranoid, you probably think this introduction is about you. In Eve Kosofsky Sedgwick, ed., *Novel gazing: Queer readings in fiction,* 1–37. Durham, NC: Duke University Press.

Sedgwick, Eve Kosofsky. 2007. Melanie Klein and the difference affect makes. *South Atlantic Quarterly* 106.3: 625–642.

Sedgwick, Eve Kosofsky, Stephen Barber, and David Clark. 2002. This piercing bouquet. An interview with Eve Kosofsky Sedgwick. In Stephen Barber and David Clark, eds., *Regarding Sedgwick: Essays on queer culture and critical theory,* 243–262. New York: Routledge.

Segal, Hanna. 1979. *Klein.* London: Karnac.

Serres, Michel. 2007. *The parasite.* Translated by Lawrence R. Schehr. Minneapolis: University of Minnesota Press.

Shapiro, Arthur, and Elaine Shapiro. 1997. *The powerful placebo: From ancient priest to modern physician.* Baltimore: Johns Hopkins University Press.

Shapiro, Jennifer, Nancy Berkman, Kimberly Brownley, Jan Sedway, Kathleen Lohr, and Cynthia Bulik. 2007. Bulimia nervosa treatment: A systematic review of randomized controlled trials. *International Journal of Eating Disorders* 40.4: 321–336.

Shell, Renee. 2001. Antidepressant prescribing practices of nurse practitioners. *Nurse Practitioner* 26.7: 42–47.

Shorter, Edward. 2011. A brief history of placebos and clinical trials in psychiatry. *Canadian Journal of Psychiatry* 56.4: 193–197.

Singh, Ilina, and Nikolas Rose. 2006. Neuro-forum: An introduction. *BioSocieties* 1.1: 97–102.

Smith, Barbara Herrnstein. 1988. *Contingencies of value: Alternative perspectives for critical theory*. Cambridge, MA: Harvard University Press.

Spillius, Elizabeth Bott, Jane Milton, Penelope Garvey, Cyril Couve, and Deborah Steiner. 2011. *The new dictionary of Kleinian thought*. London: Routledge.

Squier, Susan. 2004. *Liminal lives: Imagining the human at the frontiers of biomedicine*. Durham, NC: Duke University Press.

Stanton, Martin. 1991. *Sándor Ferenczi: Reconsidering active intervention*. Northvale, NJ: Jason Aronson.

Steiner, Riccardo. 1991. Background to the scientific controversies. In *The Freud-Klein Controversies, 1941–1945*, 227–263. London: Routledge.

Sternbach, Harvey. 1991. The serotonin syndrome. *American Journal of Psychiatry* 148.6: 705–713.

Stonebridge, Lyndsey, and John Phillips, eds. 1998. *Reading Melanie Klein*. London: Routledge.

Sulloway, Frank. 1979. *Freud, biologist of the mind: Beyond the psychoanalytic legend*. New York: Basic Books.

Symon, David, and George Russell. 1986. Abdominal migraine: A childhood syndrome defined. *Cephalalgia* 6.4: 223–228.

Szabo, Steven, Todd Gould, and Husseini Manji. 2009. Neurotransmitters, receptors, signal transduction, and second messengers in psychiatric disorders. In Alan Schatzberg and Charles Nemeroff, eds., *The American psychiatric publishing textbook of psychopharmacology*, 4th ed., 3–58. Arlington, VA: American Psychiatric Publishing.

Szekacs-Weisz, Judit, and Tom Keve, eds. 2012. *Ferenczi for our time: Theory and practice*. London: Karnac.

Teicher, Martin, Carol Glod, and Jonathan Cole. 1990. Emergence of intense suicidal preoccupation during fluoxetine treatment. *American Journal of Psychiatry* 147.2: 207–210.

Thase, Michael, Joel Greenhouse, Ellen Frank, Charles Reynolds, Paul Pilkonis, Katharine Hurley, Victoria Grochocinski, et al. 1997. Treatment of major depression with psychotherapy or psychotherapy-pharmacotherapy combinations. *Archives of General Psychiatry* 54.11: 1009–1015.

Thurschwell, Pamela. 1999. Ferenczi's dangerous proximities: Telepathy, psychosis, and the real event. *differences: A Journal of Feminist Cultural Studies* 11.1: 150–178.

Tiefer, Leonore. 2010. Beyond the medical model of women's sexual problems: A campaign to resist the promotion of "female sexual dysfunction." *Sexual and Relationship Therapy* 25.2: 197–205.

Tomkins, Silvan. 1963. *Affect, imagery, consciousness*. Vol. 2: *The negative affects*. New York: Springer.

Tomkins, Silvan. 1991. *Affect, imagery, consciousness*. Vol. 3: *The negative affects: Fear and Anger*. New York: Springer.

Trivelli, Elena. 2014. Depression, performativity and the conflicted body: An auto-ethnography of self-medication. *Subjectivity* 7.2: 151–170.

Tronick, Edward. 1989. Emotions and emotional communication in infants. *American Psychologist* 44.2: 112–119.

Turner, Judith, Richard Deyo, John Loeser, Michael Von Korff, and Wilbert Fordyce. 1994. The importance of placebo effects in pain treatment and research. *Journal of the American Medical Association* 271.20: 1609–1614.

US Food and Drug Administration. 2004. *Transcript of February 2, 2004, meeting of the Pediatric Subcommittee of the Anti-Infective Drugs Advisory Committee*. http://www .fda.gov/ohrms/dockets/ac/04/transcripts/4006T1.htm.

US Food and Drug Administration. 2007. Revisions to product label. http://www.fda .gov/downloads/drugs/drugsafety/informationbydrugclass/ucm173233.pdf.

Ussher, Jane. 2010. Are we medicalizing women's misery? A critical review of women's higher rates of reported depression. *Feminism and Psychology* 20.1: 9–35.

Waldby, Catherine, and Robert Mitchell. 2006. *Tissue economies: Blood, organs, and cell lines in late capitalism*. Durham, NC: Duke University Press.

Walsh, Timothy, Stuart Seidman, Robyn Sysko, and Madelyn Gould. 2002. Placebo response in studies of major depression: Variable, substantial, and growing. *Journal of the American Medical Association* 287.14: 1840–1847.

Watanabe, Masahito, Kentaro Maemura, Kiyoto Kanbara, Takumi Tamayama, and Hana Hayasaki. 2002. GABA and GABA receptors in the central nervous system and other organs. *International Review of Cytology* 213: 1–47.

Wheeler, Benedict, David Gunnell, Chris Metcalfe, Peter Stephens, and Richard M. Martin. 2008. The population impact on incidence of suicide and non-fatal self harm of regulatory action against the use of selective serotonin reuptake inhibitors in under 18s in the United Kingdom: Ecological study. *British Medical Journal* 336.7643: 542–545.

Whittington, Craig J., Tim Kendall, Peter Fonagy, David Cottrell, Andrew Cotgrove, and Ellen Boddington. 2004. Selective serotonin reuptake inhibitors in childhood depression: Systematic review of published versus unpublished data. *Lancet* 363.9418: 1341–1345.

Wiegman, Robyn. 2004. Dear Ian. *Duke Journal of Gender, Law, and Policy* 11: 93–120.

Wiegman, Robyn. 2014. The times we're in: Queer feminist criticism and the reparative "turn." *Feminist Theory* 15.1: 4–25.

Wilkinson, Grant. 2001. Pharmacokinetics: The dynamics of drug absorption, distribution, and elimination. In Joel Hardman and Lee Limbird, eds., *Goodman and Gilman's "The Pharmacological basis of therapeutics,"* 10th ed., 3–30. New York: McGraw-Hill.

Wilson, Edward O. 1998. *Consilience: The unity of knowledge*. London: Little, Brown.

Wilson, Elizabeth A. 1998. *Neural geographies: Feminism and the microstructure of cognition*. New York: Routledge.

Wilson, Elizabeth A. 2004. *Psychosomatic: Feminism and the neurological body*. Durham, NC: Duke University Press.

Wilson, Elizabeth A. 2011. Another neurological scene. *History of the Present* 1.2: 149–169.

Winnicott, Donald. 1949. Hate in the counter-transference. *International Journal of Psychoanalysis* 30.2: 69–74.

Wolff, Harold, and Eugene DuBois. 1946. The use of placebos in therapy. *New York State Journal of Medicine* 46: 1718–1727.

Wood, B. S. B., and Roy Astley. 1952. Vomiting of uncertain origin in young infants. *Archives of Disease in Childhood* 27.136: 562–568.

Worell, Judith, ed. 2001. *Encyclopedia of women and gender: Sex similarities and differences and the impact of society on gender*. Vol. 1. Amsterdam: Elsevier.

Wyer, Mary, Mary Barbercheck, Donna Cookmeyer, Hatice Örün Öztürk, and Marta L. Wayne, eds. 2013. *Women, science, and technology: A reader in feminist science studies*. New York: Routledge.

Zahajszky, Janos, Jerrold F. Rosenbaum, and Gary D. Tollefson. 2009. Fluoxetine. In Alan Schatzberg and Charles Nemeroff, eds., *Essentials of clinical psychopharmacology*, 4th ed., 289–361. Arlington, VA: American Psychiatric Publishing.

Zetzel, Elizabeth. 1956. Current concepts of transference. *International Journal of Psychoanalysis* 37.4–5: 369–376.

Zeul, Mechthild. 1998. Notes on Ferenczi's theory of femininity. *International Forum of Psychoanalysis* 7.4: 215–223.

Zhu, April J., and B. Timothy Walsh. 2002. Pharmacologic treatment of eating disorders. *Canadian Journal of Psychiatry* 47.3: 227–234.

Zita, Jacquelyn. 1998. *Body talk: Philosophical reflections on sex and gender*. New York: Columbia University Press.

Zito, Julie, Daniel Safer, James Gardner, Laurence Magder, Karen Soeken, Myde Boles, Frances Lynch, et al. 2003. Psychotropic practice patterns for youth: A 10-year perspective. *Archives of pediatrics and adolescent medicine* 157.1: 17–25.

INDEX

abdominal migraine, 13–16

Abraham, Karl, 7, 69, 75–78, 182n7, 185n5, 189n8

abreaction, 73–74, 84, 87–88

adverse effects of antidepressants, 11, 100, 116, 124, 142–145, 149–166

aggression and feminist theory, 1–2, 5–6, 16, 23, 41, 67–75, 82–93, 149–151, 166–167

ambivalence, 69–70, 73, 77, 88, 189n6

amphimixis, 35, 61–62, 65–66, 112, 140, 172

analytic third, 110–112. *See also* Odgen, Thomas

anatomy and feminist theory, 26, 35, 45–49

Angell, Marcia, 117, 150, 195

anger, 5–7, 63, 65, 68–69, 72–74, 83, 87, 89, 107, 151, 157, 160, 187n2

animism, 38, 43, 182n5, 182n7

anorexia, 52, 59, 75, 186n8

antibiologism and feminist theory, 1–5, 16, 24–26, 30–34, 43, 66, 93, 171–172

antidepressants, 21–22, 93, 112–117; bulimia and, 63–67; chemical characteristics of, 98–104; feminism

and, 11–12, 129; monoamine oxidase inhibitors (MAOIs), 11, 63, 125, 127, 153–154, 161; prescription of, 7–12, 149–151, 174; tricyclic, 7, 11, 63, 125, 127, 191n1. *See also* pharmakon; placebo; randomized controlled trials of antidepressants (RCTs); selective serotonin reuptake inhibitors (SSRIs); suicidal ideation

antipsychiatry, 10–12, 154; deinstitutionalization and, 9

antisocial thesis in queer theory, 56

anxiety, 26, 59, 88, 138, 152, 162, 170–177

Aron, Lewis, 110, 185n4

Austin, J. L, 76, 164

Barad, Karen, 9, 13, 66, 112, 133, 145, 198n2

Beck, Aaron, 7

Beck Depression Inventory, 157

Beecher, Henry, 139

belly, as minded, 24, 27–28, 31–32, 35, 38–40, 43. *See also* Klein, Melanie; Rubin, Gayle; stomach

Berlant, Lauren, 85; and Lee Edelman, 6, 70, 169, 176, 178, 189n5

Bersani, Leo, 5–6; and Melanie Klein, 88, 182n4, 190n10

Beyond the Pleasure Principle, 79, 83

Big Pharma, 97, 106, 117, 120, 129, 150, 154, 182n6

bile, 1, 5, 7, 16, 70. *See also* bitter

bingeing, 59, 62–64. *See also* bulimia

biological phantasy, 36–43, 172. *See also* Isaacs, Susan; Klein, Melanie

biological unconscious, 23, 40–41, 44, 49, 55–58, 64, 70. *See also* Ferenczi, Sándor

biology: feminist theory and, 3–5, 13, 29–35, 49–51, 58; as flat, 56–59, 61–62, 64–66, 70, 77, 81, 83, 173; as juridical, 34–35, 38, 43, 59, 172

Birke, Lynda, 3

bitter, 70, 75, 81, 85, 93, 178. *See also* bile

black box warnings, 151, 198n3. *See also* Food and Drug Administration (FDA)

blood-brain barrier, 101, 103–104, 117, 156

Bordo, Susan, 184n2

bowel, 1, 13, 52, 62, 76–78, 101

brain-derived neurotrophic factor (BDNF), 162

Breggin, Peter, and Ginger Breggin, 10, 98, 192n5

Brierley, Marjorie, 36–38

British Psycho-Analytical Society. *See* controversial discussions

bulimia, 49, 59–63, 186nn8–9; antidepressants and, 63–67. *See also* bingeing

Butler, Judith, 23, 29, 69, 190n10; aggression and, 88–89; gay melancholia, 87–88; gender melancholy, 86–87

catharsis. *See* abreaction

Celexa, 101–102, 126, 149. *See also* selective serotonin reuptake inhibitors (SSRIs)

Chesler, Phyllis, 68

Cixous, Hélène, and Catherine Clement, 73

clinical trials of antidepressants. *See* randomized controlled trials of antidepressants (RCTs)

cognitive behavioral therapy (CBT), 146–147, 150

Committee on Safety of Medicines (CSM), 155, 158

consilience, 27, 106, 112, 170, 172, 175–176

controversial discussions, 36–39

Corbett, Ken, 85, 90–92

Crimp, Douglas, 85, 86

Cvetkovich, Ann, 10, 85–86, 97, 127, 154, 192n2

Davis, Lennard, 10, 97, 106, 127, 154, 199n10

death drive, 41, 84, 190n10, 190n13

depression: as anger turned inward, 69, 72, 88, 187n2; biological markers of, 133–136, 162; in children and adolescents, 16, 79, 92–93, 141–142, 148–149, 151–157, 162–166, 199n8; cognitive theories of, 7, 187n1; history of, 7–8; neurological theory of, 162–165; pediatric, 11, 13, 141, 146, 149–159, 162–166, 198n5; pharmaceutical treatment of, 2, 9, 11–12, 16–17, 21–22, 63, 97, 100, 104, 113–120, 128, 143–146, 162–165, 170, 182n6, 195n15; psychoanalytic theories of, 7, 22, 68–70, 71–78; women and, 7–8, 11–12, 59, 68, 97, 187n1, 188n4, 189n7. *See also* hostility; "Mourning and Melancholia"; suicidal ideation

Derrida, Jacques, 141–145. *See also* pharmakon

Deutsch, Felix, 57–58

nervous system, central (CNS), 5, 12, 14–15, 66, 98–104, 142, 156–157, 161, 164, 172–174

nervous system, enteric, 5, 66

Neuropsychiatric Institute, UCLA, 133–134, 136, 139

neuroscience, 30, 146, 162; feminist theory and, 4, 5, 28–29, 149, 171, 175; humanities and, 4, 12, 27, 35, 49, 59, 164, 171–176

object relations, 39, 40, 78

Ogden, Thomas, 110–115, 119, 194nn9–11

organ speech, 75–76, 79

Oyama, Susan, 9, 66, 188n4

paranoid reading, 4, 70, 71, 177–178, 181

parasitism, 124, 132–133, 135, 139

Paxil, 101–103, 126, 149, 197n6. See also selective serotonin reuptake inhibitors (SSRIs)

Pepper, Oliver, 121–123

peristalsis, 22, 61–62, 78

persistent depressive disorder. See dysthymia

Persson, Asha, 144–145

phantasy, 21–22, 27, 90, 93, 147, 179, 194n11; biology and, 35–44, 171–172, 176; definition of, 181n1; Ferenczi and, 50–52, 65–67; Klein and, 71, 76–78, 182n7, 183n8, 189n6; transference and, 109, 111

pharmakon, 141–146, 165

pharynx, 61–62, 80

phylogenesis, 41, 53–55, 58–62, 185n5

placebo, 11, 16, 63, 67, 93; in comparison to antidepressants, 120–142, 149–151, 156; EEG measure of, 134–136, 140, 163. See also random-

ized controlled trials of antidepressants (RCTs)

Prozac, 7, 9, 11–12, 64, 74, 97, 100–103, 115–118, 126, 149, 153–154, 186n9, 191nn1–3, 192n5, 194n12. See also selective serotonin reuptake inhibitors (SSRIs)

psychobiotics, 169–173

psychosis, 51, 55, 107–108, 110, 122, 165–166, 184n3, 193n8, 199n11

psychosomatic. See mind-body relationship

randomized controlled trials of antidepressants (RCTs), 63, 122–128, 130–139, 148, 151, 155–160, 196n5, 199n8

reparative reading, 70–71, 176–178. See also paranoid reading

repressive hypothesis, 33–34

Rome III guidelines for functional gastrointestinal disorders, 80–81

Rose, Jacqueline, 37, 41, 182n4, 183n8, 190n10

Rubin, Gayle, 23–35, 40, 43, 58, 184n2

rumination disorder. See merycism

Russell, Gerald, 14, 15, 60–62, 64

sadism, 69, 70–75, 77–78, 85, 87, 93, 182n7, 188n3, 190n13

Salamon, Gayle, 90–92, 187, 189n7

Schiesari, Juliana, 187n1, 188n4, 189n7

Sedgwick, Eve Kosofsky, 38, 70–71, 177–178, 189n5

selective serotonin reuptake inhibitors (SSRIs), 9–11, 63, 100, 112–117, 119, 142, 144, 146, 153, 161–163, 186n9, 191n1, 192n3, 197n6; bioavailability of, 101–102, 192n4; half-life of, 102; lock and key model of, 104–105, 193n6; metabolites of, 98–99, 102–103; oral administration of,

CPI Antony Rowe
Eastbourne, UK
November 30, 2022